An Agent-Based Model of Heterogeneous Demand

Matthias Müller

An Agent-Based Model of Heterogeneous Demand

With a foreword by Prof. Dr. Andreas Pyka

 Springer VS

Matthias Müller
Stuttgart, Germany

Dissertation University of Hohenheim, Germany, 2016

D 100

ISBN 978-3-658-18721-7   ISBN 978-3-658-18722-4   (eBook)
DOI 10.1007/978-3-658-18722-4

Library of Congress Control Number: 2017944918

Springer VS
This Springer VS imprint is published by Springer Nature
The registered company is Springer Fachmedien Wiesbaden GmbH
The registered company address is: Abraham-Lincoln-Str. 46, 65189 Wiesbaden, Germany

Foreword

Published in 1982, probably the most quoted book in innovation economics is *An Evolutionary Theory of Economic Change* by Richard Nelson and Sidney Winter. Today, it is considered as a cornerstone for modern evolutionary economics and thirty-four years later, evolutionary economics has created numerous insights addressing economic development and innovation-driven change.

Against this backdrop, it is astonishing that the relationship between innovation and demand has been largely neglected and that innovation economics has been often limited to the analysis of supply side effects exclusively. Instead, this issue was addressed first by innovation politics, applying not only public procurement but also subsidies for consumers as an effective tool for fostering innovation. Important examples of this are the renewable energy act securing guaranteed prices above market prices for sold electricity of photovoltaic systems or subsidies for electric vehicles in Germany. It is without doubt that new innovations require consumers who are willing to spend a part of their incomes. With this we see a first important link in the co-evolutionary relationship between income and technological development. At the same time, however, we have to consider that the particular demand of heterogeneous consumers is an important key for the diffusion of innovations. The demand of consumers determines which innovations will be successful and how successful innovation processes should be structured. At an extreme, we see that consumers often also actively participate in innovation processes co-designing innovations in so-called user-producer-relations. Although the debate between a technology-push or demand-pull perspective dates back to the late 1970s and has resulted in favour of a balanced view as noted for example by Nathan Rosenberg, today, the demand side is still somewhat neglected within innovation economics. In his thesis, Matthias Müller addresses this research gap and analyses the multi-faceted interplay of demand and innovation processes especially for the sectoral development based on a theoretical model. Central to his work is an agent-based computer simulation of innovation and demand. Matthias Müller's research represents an important milestone and, at the same time, the basis for further research on this issue relevant not only from a theoretical perspective but also for the application of a new understanding of the complex relationship between innovation and demand.

Prof. Dr. Andreas Pyka

Acknowledgements

I would like to thank my first supervisor Professor Pyka for his support, advice, useful suggestions and for the freedom he granted me to work on the dissertation and my research projects beyond. I thank my colleagues Tobias Buchman, Kristina Bogner, Bianca Janic, Michael Schlaile, Benjamin Schön, Sophie Urmetzer and my co-authors Muhamed Kudic and Benjamin Schrempf for the inspiring discussions and help during the last years. Last but not least, I thank my family, my friends and Anna-Lena Brüning for their support and their patience.

Matthias Mueller

Table of Contents

Foreword...5

Acknowledgements..7

Table of Contents..9

List of Figures...11

List of Tables..15

List of Equations...17

List of Symbols...19

1   Introduction...21

   1.1   The Invention of the Wheel...21

   1.2   Methodological and Modelling Framework...................................25

   1.3   Research Question and Outline..29

2   The Role of Consumers in Innovation Economics.................................33

   2.1   The Neglected Demand Side...33

      2.1.1   The Linear Innovation Model...33

      2.1.2   The Demand-Pull Modell of Innovation.............................36

      2.1.3   A Multidimensional Perspective on Innovation Processes.......40

   2.2   An Evolutionary Perspective...43

   2.3   The Role of the Demand Side Today...49

3   The New Agent-Based Paradigm...57

   3.1   Three Pillars of ABM..57

      3.1.1   Modelling from an Agent-Based Perspective......................57

      3.1.2   A Definition of Agents..59

      3.1.3   Simulation as In-Silicio Laboratories.................................61

   3.2   Using ABM as a Scientific Tool..64

      3.2.1   Why Do We Need Agent-Based Modelling?.......................65

      3.2.2   Managing the Complexity...66

      3.2.3   Two Ways of Using Agent-Based Models...........................69

      3.2.4   The Need for Verification, Validation and Calibration...........71

   3.3   Implications for the Following Analysis...72

4   An ABM of Heterogeneous Consumers and Demand.............................75

4.1   Introducing Remarks 75

4.2   The Baseline Simulation Model 79

4.2.1   Modelling Multi-Dimensional Product Characteristics 79

4.2.2   Basic Procedure of the Simulation Model 82

4.3   Simulation Experiments 85

4.3.1   Innovations in a Multidimensional Characteristic Space 86

4.3.2   Markets in-between Homogeneous and Heterogeneous
Demand 93

4.3.3   Implications for Innovation Policies 98

4.4   Discussion 103

5   Networks of Heterogeneous Agents 105

5.1   Informal Knowledge Exchange in Firm Networks 105

5.1.1   Introducing Remarks 105

5.1.2   Knowledge Exchange and Network Formation Mechanisms . 108

5.1.3   Model Analysis 111

5.2   The Importance of Consumer Networks 125

5.2.1   Introducing Remarks 125

5.2.2   Bounded Rationality of Consumers 127

5.2.3   The Effects of Consumer Networks 131

5.3   Discussion 139

6   Bounded Morality of Consumers 143

6.1   Introducing Remarks 143

6.2   Model Analysis 145

6.2.1   Baseline Scenario 145

6.2.2   Responsible Innovation and Limited Information 150

6.2.3   Responsible Innovation and Networks 153

6.3   Discussion 156

7   Discussion and Further Research Avenues 159

Literature 163

List of Figures

Figure 1:   Smith's Circular Growth 34

Figure 2:   The Linear Innovation Model 36

Figure 3:   The Kaldorian Demand-Pull Mechanism 37

Figure 4:   The Demand Pull Model 38

Figure 5:   The Chain-Link Model 42

Figure 6:   Adopter Categorization after Rogers 50

Figure 7:   Flock of Birds Created by the BOIDS Algorithm at Different
             Points in time 58

Figure 8:   NetLogo's Graphical User Interface 62

Figure 9:   Idealised steps of Using Simulations 63

Figure 10:   The Micro and Macro Level of ABM 69

Figure 11:   Scope of the Baseline ABM 77

Figure 12:   Graphical User Interface of the Simulation Model 79

Figure 13:   Relative Distance of Demand for Different Levels of p 82

Figure 14:   Iterative Steps of the Simulation 83

Figure 15:   Average CPM Level for Homogeneous (left) and
             Heterogeneous (right) Consumers 87

Figure 16:   2-Dimensional Search Problem with Homogeneous Demand 88

Figure 17:   2-Dimensional Profit Landscape with Homogeneous Demand 89

Figure 18:   2-Dimensional Search Problem with Heterogeneous Demand 89

Figure 19:   2-Dimensional Profit Landscape with Heterogeneous Demand ... 90

Figure 20:   Innovations During a Simulation Run for Homogeneous (left)
             and Heterogeneous (right) Consumers 92

Figure 21:   Number of Firms for Different Levels of Heterogeneity 93

Figure 22:   An Indicator of Dynamic Segmentation 94

Figure 23:   Average CPM Level for Different Levels of Heterogeneity 95

Figure 24:   CPM level Distribution 96

Figure 25:   Sales Distribution for Different Levels of Heterogeneity 96

Figure 26:   Histogram of Firm Size (for different levels of p) 97

Figure 27:   Market Size, Length of Characteristic Space and the Number
             of Firms 99

Figure 28:   Market Size, Length of Characteristic Space and Average
             CPM Levels 100
Figure 29:   Effect of Sensitive Consumers 101
Figure 30:   Number of Firms for Different Policy Settings 102
Figure 31:   Networks within the Simulation Model 112
Figure 32:   Average Knowledge Levels of Agents 113
Figure 33:   Knowl. Levels of Agents for Different Absorptive Capacities ... 114
Figure 34:   Degree Distribution in the Four Networks 116
Figure 35:   Var. of Degree Distribution and Average Knowledge Levels 117
Figure 36:   Cumulative Number of Non-Traders 117
Figure 37:   Relationship Between the Time Agents Stop Trading and
             Their Degree 118
Figure 38:   Relationship Between Knowledge Levels and Degree 120
Figure 39:   Relative Position and Average Knowledge Levels 121
Figure 40:   Average Knowledge Levels of Small Agents with Policy
             Interventions 123
Figure 41:   Average Knowledge Levels of Small Agents with Policy
             Interventions 125
Figure 42:   Number of Firms (left) and CPM Levels (right) with Limited
             Consumer Information 127
Figure 43:   Innovations During a Simulation Run with Limited
             Information (left: $p = 0$, right $p = 1$) 128
Figure 44:   Reduced Segmentation in Heterogeneous Markets 129
Figure 45:   Average CPM and the Number of Informed Consumers 130
Figure 46:   Example of Consumer Networks 131
Figure 47:   Number of Firms with by Consumer Networks 133
Figure 48:   Consumer CPM Levels with Consumer Networks 134
Figure 49:   Degree Distribution of Homophily (left) and Heterophily
             (right) Consumer Networks 135
Figure 50:   CPM Levels for Homophily and Heterophily Cons. Networks .. 135
Figure 51:   Degree and CPM Levels for Different Network Topologies 136
Figure 52:   Min. CPM of Consumers in Case of Homogeneous Demand 137
Figure 53:   Relative Distance Between Consumers over Time 138

Figure 54:   Number of Firms and CPM Levels Over Time 139

Figure 55:   Match with Negative Characteristics and CPM Levels for
              Different Degrees of Consumer Heterogeneity 146

Figure 56:   Match with Neg. Characteristics in a Fully Informed Market 147

Figure 57:   CPM Levels in a Fully Informed Market 148

Figure 58:   Match with Neg. Characteristics of Responsible Consumers 148

Figure 59:   Match with Negative Characteristics of Normal Consumers 149

Figure 60:   Sensitivity of Fully Rational Consumers 150

Figure 61:   Match with Negative Characteristics with Limited Information. 151

Figure 62:   Number of Firms in Case of Limited Information 152

Figure 63:   Match with Negative Characteristics of Normal Consumers
              with Limited Information 152

Figure 64:   Sensitivity of Boundedly Rational Consumers 153

Figure 65:   Match with Negative Characteristics with Networks 154

Figure 66:   Match with Negative Characteristics of Normal Consumers
              with Networks 154

Figure 67:   Match with Negative Characteristics for Different Network
              Topologies 155

Figure 68:   Number of Firms for Different Network Topologies 156

List of Tables

Table 1: Products' Quality Values 53

Table 2: Parameters and Initial Values of the First Experiment.................. 86

Table 3: Average Path Length and Cliquishness of all Four Network
Topologies 114

Table 4: Network Characteristics of all Four Network Topologies 132

Table 5: Network Char. of Homophily and Heterophily Networks. 134

List of Equations

(1) Kene Structure.. 52

(2) Example of a Kene.. 52

(3) Innovation Hypothesis 53

(4) Definition of a Product... 80

(5) Knowledge Units of Firms 80

(6) Knowledge Mapping Functions 80

(7) Relevant Information for a Product Characteristic......................... 81

(8) An Example of a Product Characteristic 81

(9) The Individual Demand... 81

(10) Example of a Knowledge Stock 83

(11) Product Category Mapping Function 84

(12) Temporally Product Category 84

(13) Knowledge of Firms.. 84

(14) Product Characteritsic 84

(15) Inverted Hamming Distance... 84

(16) The CPM Level 85

List of Symbols

Symbol	Description
$\overline{K}_j$	firm j's knowledge stock
$K_{i,j}$	knowledge unit K_i of firm j
$C_{1,\ldots,n}$	firms' capabilities
$A_{1,\ldots,n}$	firms' abilities
$E_{1,\ldots,n}$	firms' expertise levels
IH_j	knowledge pieces of firms unused in a particular production process
v_X^i	measure of product X in respect of characteristic i
$A_{k,j}$	set of characteristics of product k of firm j
$M_{A_{k,j}}$	mapping function of product k
$C_{k,j}$	product characteristics of product k of firm j
$D_{k,n}$	demand of consumer n for product k
p	global parameter to determine the level of heterogeneity of demand in the simulation
$\Delta(A_{k,j}, D_{k,n})$	Hamming distance between the set of characteristics of product k and the demand of consumer n for this product k
φ_j	firm j's market share
h^r	threshold for radical research
h^i	threshold for incremental research
s_N	sensitivity of consumer
$v_i = (v_{i,c})$	knowledge vector of agent i with knowledge categories c
$\overline{v}$	knowledge stock of all agents over time in the barter trade model

$P(k) \sim k^{-\gamma}$	probability $P(k)$ that a node in the network is linked with k other nodes
w	probability that links within a regular ring network lattice are randomly redistributed
α_i	absorptive capacities of agent i
ω_i	an agent i's realitive position in the network indicated by the degree difference between i and his direct partners
ε	number of randomly chosen informed consumers
γ	individual importance of product characteristics
δ	ratio between consumers not considering negative product characteristics and responsible consumers

1 Introduction

1.1 The Invention of the Wheel

Today, the word innovation is something everybody knows. It is one of those buzz words which one encounters in many different ways and occasions. Firms advertise even slight changes as brand new innovations and are never tired of emphasising their innovative behaviour. Even politicians recognise the importance of at least mentioning *innovation*, and policies to stimulate innovation have become an important topic for the government. Also the European Commission, for example in its *Innovation Union – A Europe 2020 initiative*, calls for innovation as a central element in its attempt to stimulate the European economy:

> "We are facing a situation of 'innovation emergency'. […]. Thousands of our best researchers and innovators have moved to countries where conditions are more favourable. Although the EU market is the largest in the world, it remains fragmented and not innovation-friendly enough." (European Commision 2016)

Facing this claim, it is remarkable how little we know about innovation and how and why they are created. Until the early 19th century, economic growth was believed to be achieved solely via an increased use of production factors such as capital or labour. Due to the seminal work of, for example, Schumpeter (1912, 1942) and Robert Solow (1956, 1957) this view has changed today and innovation perceived as technological progress has been identified as one of the main forces that drives economic growth and prosperity. But do we really understand the innovation process in all its complexity?

Looking back into human history, it becomes clear that innovations shape human life in every imaginable way. Starting with the invention of the wheel, which can be dated back to the early Bronze Age, the list of substantial innovations seems endless. Animal powered ploughs, the steam engine, the combustion engine, electric power, the first airplanes, and modern communication devices such as smartphones are just a few examples of important milestones in human development. Each of these represents a substantial change not only for the economy but also for every man's life.

Determining the most important invention of all time may sound impossible. Putting forth the ideas of Alfred North Whitehead, the answer may also be simple: it is the method of the invention itself. What Whitehead suggests in his famous book from 1925 *Science and the modern world* about the inventions of the nineteenth century, is a fascinating and significant insight into the heart of the innovation process. Instead of evaluating the impacts or the alleged complexity of different innovations, Whitehead put forward the idea that the way innovations are created changed by the end of the 19th century. He states that the process of inventing before this transformation was slow, unconscious, and unexpected, and it became quick, conscious and expected (Whitehead 1975, p. 120).

What Whitehead hereby acknowledges is nothing less than the inherent complexity of the invention process (and with that: of the innovation process) and the revolution of how we deal with it. Although in retrospect some innovations may seem to be the result of single Eureka moments of single inventors, it would be a false conclusion that innovation takes place in an isolated world. The wheel for example, ranking first in the above list of inventions, was not simply the humble result of a single inventor. It was one of the first inventions that was not inspired by the natural world (LaBarbera 1983) and was the result of the complex interplay of different technologies, knowledge, and other factors of that time, although it is sometimes unjustly seen as simple and trivial, today.

Surprisingly, the first records of wheels in use for transportation purposes show that wheels weren't invented until 3,500 a. Chr., which is relatively late in human history, considering the technological capabilities of that time and other prominent inventions dating before the wheel. One clear reason why the wheel was invented so late in human history is that metal tools were needed to achieve the accuracy to guarantee a smooth and frictionless combination of wheels and axles. Using stone tools to shape perfect circular wheels and drill holes for the axles would have been a hassle. Producing the right metal tools such as drills, saws, and grinding tools, however, requires the knowledge and the technology to shape and make use of metal that meets the requirements for that purpose. This in turn requires an expertise in identifying the right raw materials, i.e. stones containing metal ores, producing coal, ovens to achieve the right temperature, and eventually the knowledge of how to combine the right ores to produce bronze. In other words, the wheel was not invented earlier because it was built from a large set of knowledge and previous inventions. In fact, with this background, the invention of the wheel appears incredible for its time.

Today, the complexity behind the innovation process, the interplay of different technologies, knowledge and the versatility of actors involved is increasingly recognized, not only in the broad field of economic science, but also by politicians and policy instruments. Without this *new art of invention* of the 19th century, as Whitehead puts it, modern inventions were simply impossible. What changed during the 19th century was the way science and technology was perceived and how the people of that time managed the complexity of the scientific process. This lead to a new and ground-breaking professionalism in science and the broad establishment of universities and other research institutes designed to systematically create new scientific knowledge and methods and thereby made complex inventions possible (Whitehead 1975).

There is, however, a second reason for the late invention of the wheel and this reason is easily neglected. We have to consider the increasing demand for the efficient transportation of goods and commodities at that time. While the back of an animal such as a horse, donkey or an ox may be sufficient for bridging short distances, increasing trade of goods around the world and thus intensifying the establishment of common trade routes created an enormous demand for that invention for the first time in human history. Second, the story would be incomplete if we neglected an additional factor which triggered the invention of the wheel as a means for transportation. In fact, we have to consider that the invention of the wheel as a means for transportation was not the first invention of the wheel. Centuries prior to that, simple forms of wheels appeared in pottery, enabling potters to easily produce simple but effective containers to carry water, nutrition, etc. Driven by the large demand for these products, the technology of producing pottery wheels quickly diffused and at some point was improved so that wheels for transportation were also possible. In other words, one might speculate that the demand for pottery created the necessary basic prerequisite for the invention of the wheel.

The demand side, e.g. users and consumers, is an important element in the whole picture of the innovation process. Unfortunately, it is too often neglected or in the best case oversimplified in economics. Admittedly, it would be false to state that the demand side and the role of consumers have not been considered at all. Instead, we have to be more precise and ask *how* it has been considered as an element of the innovation process.

Until the middle of the 20th century the focus of the scientific discourse was on the question of whether the demand side has effects on the innovation process. Today, after a fruitful discussion in the literature, the debate branched into several aspects dealing with the question of how the demand side influences the innovation process. The well-known concept of user-innovation by Eric von Hippel (1976, 1988) is just one example in which consumers are appreciated as important actors in the innovation process. Following this idea, consumers sometimes act as innovators themselves, creating novel solutions for the particular and individual needs they have. However, by far the larger portion of work on innovation and technological change is concentrated on supply-side dynamics (see for example Adner, Levinthal 2001, Coombs 2001, Witt 2001a, Harvey et al. 2001, Andersen 2007, Ciarli et al. 2008, Nelson, Consoli 2010).

One possible explanation for this fact lies in the origin of modern innovation economics theory. Innovation as an economic concept can be traced back to the work of Joseph Alois Schumpeter, who stands with his work in-between the work of Walras and Marx (Kurz 2005). Schumpeter followed with his theory some of the ideas of Walras and hereby adopted the perspective that the total demand for goods and services will always adjust to (or be equal to) total supply (Knell 2012). In this perspective, innovations are created because entrepreneurs push innovation, rather than because consumer needs pull them.

This opinion was challenged by the seminal work of Schmookler (1962, 1966) and other studies, starting an ongoing debate between demand-pull arguments on the one side and technology-push arguments on the other side. Unsatisfied with the linear view of a supply-oriented focus on the innovation process, Schmookler (1962, 1966) shows by studying different industries of the US economy that peaks in patenting activities lag behind peaks in the production of commodities. He concluded that the influence (upon innovation) of the latter (unfolding economic needs) has been substantial, at least in established industries (Schmookler 1962, p. 20). This understanding of the demand side also fits beautifully with a famous quote by Ludwig von Mises (1949, p. 269):

> "The direction of all economic affairs is in the market society a task of the entrepreneurs. Theirs is the control of production. They are at the helm and steer the ship. A superficial observer would believe that they are supreme. But they are not. They are bound to obey unconditionally the captain's orders. The captain is the consumer."

Since the first impetus by Schmookler in 1969, a series of empirical studies have tried to support his hypothesis (Andersen 2007) while others viewed the activities and internal capabilities of firms as the primary drivers of innovation (Teece 1986). As a preliminary result of the debate between demand and supply arguments, both the demand and the supply side appear to simultaneously play crucial roles (Mowery, Rosenberg 1979, Nelson, Winter 1982). Successful innovation emerges from the interaction between demand-pull and technology-push effects (Mowery, Rosenberg 1979). However, the spirit of the original debate is also expressed in the current debate on demand-side innovation policies which aim to foster innovation and economic development by measures focusing on the demand side rather than on the supply side (see Edler 2007, Edquist 1994 for an overview and discussion on that issue). In more detail, today we find a large set of different policy instruments aiming at the demand side for fostering innovation activities such as public procurement, financial subsidies, regulation and finally information and enabling (Edler, Georghiou 2007, Edler 2009). Green energy supply, E-mobility are just some examples of potential fields for which demand-sided innovation policies are currently discussed and at least partially applied.

Although at the very heart of this debate the demand side was identified as an important factor for the innovation process, the way consumers and their needs are included in the analysis often lacks important details. If we want to understand the innovation process and the interactions between the demand side and the supply side in more detail, we need to acknowledge both the actors behind the demand side and the supply side as what they are: multifaceted and heterogeneous entities embodied within the complex system of our economy. As a consequence, understanding innovation processes in detail requires a suitable and appropriate theoretical but also methodological framework.

The analysis in this dissertation builds on the cornerstones of the so-called evolutionary economics approach in a neo-Schumpeterian fashion which in contrast to the neoclassical school of economic thought allows for a more detailed picture of the complex system behind innovations.

1.2 Methodological and Modelling Framework

As with most economic theories it is hard to define absolute criteria of what the neoclassical approach is and which set of assumptions it includes (see also Nelson, Winter 1974 for a discussion on the difference between neoclassical and evolutionary theories). Following Hodgson (1998a, p. 169), however, the core characteristics of the neoclassical schools can be defined as it:

> "(1) assumes rational, maximizing behaviour by agents with given and stable preference functions, (2) focuses on attained, or movements toward, equilibrium states, and (3) excludes chronic information problems […]".

This set of assumptions guarantees an analytical traceability, which in turn allows for simple models with explicit implications. The detailed analysis of innovation, as also the previous section revealed, however, cannot be built on such restrictive and mechanic assumptions.

To stress this issue further, let us briefly consider the following points. At the very core of any innovation stands that it upsets every equilibria (Schumpeter 1912, Schumpeter 1942, Nelson, Winter 1982). In fact, as Antonelli and Scellato (2008, p. 2) put it: "Innovation is not only the cause of out-of-equilibrium conditions, but also the consequence of out-of-equilibrium". As a consequence and also as Schumpeter argues, it is not enough to look through a static lens, instead, economic development has to be seen as a process of qualitative change driven by innovation and occurring in historical time (Fagerberg 2004, p. 6). The analysis of innovation from an equilibrium oriented framework, thus, would be inappropriate in the best case.

An approach, which gives us the possibility to take a dynamic perspective is the so-called evolutionary economics approach (Witt 1993). The dynamic perspective signifies a movement of something over time or as Dosi and Nelson (1994, p. 328) put it: "to explain why that something is what it is at a moment in time in terms of how it got there". With this, evolutionary economics focuses on the analysis of *far-from-equilibrium* processes occurring in complex systems, which show close and frequent interactions of components (see for example Hanusch, Pyka 2007a).

Second, innovations are the result of complex intentional but also unintentional decisions of heterogeneous entities facing limited capabilities and information (Nelson, Winter 1982, Dosi, Nelson 1994, Kwasnicki 2007). Assuming that either firms or consumers show rational, maximizing behaviour with given and stable preference functions without any information problems simply neglects the sheer reality of any economic system. So instead of assuming perfect rational representative agents, as neoclassical theories tend to, we need to take all economic actors as they are: individual entities embedded in an economic system.[1] Although this discussion can be extended in various ways (see also Kudic 2014 for a discussion on this issue), summing up, it becomes clear that for a thorough understanding of the role of the demand side we need to build our analysis on a framework which avoids such restrictive assumptions.

Evolutionary models of demand consider the main features of the evolutionary theory. Based on this we can characterise these models firstly as *dynamic*. Dynamic in this sense means a dynamical, spontaneous and historical process in which macroeconomic characteristics are the result of activities observed at the micro-level (Nelson, Winter 1982, Dosi, Nelson 1994, Kwasnicki 2007). Second, and strongly related to the first issue, the focus of these models lies on the analysis of *far-from-equilibrium* processes in complex systems which show close and frequent interactions of components (see for example Hanusch, Pyka 2007a). Furthermore, these models need to deal with the *heterogeneity* of economic agents (Kwasnicki 2007). Heterogeneity in the course of the evolutionary model of demand, however, must include the individual characteristics and behaviour of all economic actors and, hence, also include the demand side. Finally, evolutionary models include the concept of *bounded rationality*. While this concept is usually only applied on the firm side, without doubt, also consumers are far away from an Olympic rationality in their decision-making. Consumers cannot know about the properties of all goods on the market because of individual limits of information, knowledge or effort (Faber, Frenken 2009).

Within the evolutionary literature, analyses of demand are still at an early stage (Ciarli et al. 2008). The vast majority of evolutionary models still

---

1   In fact, the most common justification for the perfect rational representative agent is the *as if* argument by Milton Friedman (1953), however, this argument can be applied only to agents subject to competitive pressure, which does not apply for consumers (Valente 2009).

oversimplifies the demand side in the tradition of Schumpeter's ideas. As for example Nelson and Consoli (2010, p. 667) note:

> "evolutionary economics badly needs a behavioral theory of household consumption behavior, but to date only limited progress has been made on that front. Partly because Schumpeter's own writings were focused there, and partly because this has been the focus of most of the more recent empirical work on technological change, the lion's share of the writing by modern evolutionary economists concerned with economic dynamics has been focused on the 'supply side' of economic behavior: the behavior of firms, the nature of innovation and technological progress, industrial competition and dynamics"

Nevertheless, the question remains why, as Nelson and Consoli phrase it, the most of the recent empirical work on technological change has been focused on the supply side of economic behaviour leaving an important theoretical gap. Given the wide evidence about the importance of the demand side and the potential implications, a general lack of interest is rather unlikely. Additionally, it would be too easy to assume that most scholars simply follow the tradition of Schumpeter. Which brings us to the point that yet another, additional characteristic of innovations may be responsible: the complexity involved.

In fact, the theoretical gap cannot easily be bridged by traditional modelling approaches and their limited framework to analyse the behaviour of complex systems. Extending the focus on heterogeneous and boundedly rational consumers calls for new modelling approaches suited to analyse the complex interplay between economic actors. Since the 1980s, continuous improvements in computer technologies have remarkably changed scientists' possibilities to develop and apply computer simulations. Simulations as a scientific tool offer new ways to explore the dynamics of complex models in various disciplines. They enable us to process scientists' thought models into artificial test laboratories in which we can systematically analyse the models' outcomes.

Within the broad field of simulation techniques, the so-called agent-based modelling (henceforth ABM) approach has gained increasing momentum, not only for economics but also in many other scientific disciplines. The ABM approach takes the perspective of the system building elements and focuses on the actions and interactions of these entities as the relevant actors within the system. This perspective of ABM, is accompanied by the attempt to represent actors of economic systems in a more realistic fashion, overcoming the shortcomings of approaches limited to representative agents which by definition ignore heterogeneity and the related implications of interacting heterogeneous agents.

The value of such a computational modelling approach becomes clear when we look at the experience gained during the 2010 financial crisis. As Jean-Claude Trichet phrased it:

"When the crisis came, the serious limitations of existing economic and financial models immediately became apparent. [...] Macro models failed to predict the crisis and seemed incapable of explaining what was happening to the economy in a convincing manner. [...] We need to deal better with heterogeneity across agents and the interaction among those heterogeneous agents. [...] Agent-based modelling dispenses with the optimization assumption and allows for a more complex interaction between agents. Such approaches are worthy of our attention."[2]

Given the enormous degree of complexity within the innovation process we need a methodological framework, such as ABM, to incorporate the individual and heterogeneous behaviour of economic actors into models which are capable to portray the emerging dynamics. Despite the new and promising perspective ABM offers, it is also necessary to deal with the question of how ABM can contribute to the scientific endeavour. If used properly, ABM offers researchers both a new scientific method and a new perspective on the complex interplay between actors within an economic system.

Nevertheless, it is too easy to say we simply include all possible aspects and elements of the picture. Murray Gell-Mann (1995) has been attributed with saying: "Imagine how hard physics would be if electrons could think". Economic science has to face exactly this problem. The subject under investigation in economics is something human. Humans by definition are not homogeneous nor are they fully rational as some schools of economic thought assume.[3]

Shifting the focus from an oversimplifying perception of economic systems to a more realistic one, ABM is designed to overcome the limiting possibilities of a traditional analytical framework. Central to this new approach is its exceptional perspective of economic actors, treating economic agents as heterogeneous and individual actors which build economic systems by their decisions, actions and interactions from the bottom up.

However, using this method comes at cost. We need to be aware of how ABM can be used to deepen our understanding of complex economic processes and how models should be designed. Although ABM at its core makes a huge step towards a more realistic model of economic systems, one cannot expect a fully detailed picture. As with any model, an ABM is designed as a purposeful representation of a system. Purposeful in that sense means it can be used in a generative way, reproducing macro level pattern and finding possible explanations on the micro

---

2   Quoted from the speech 'Reflections on the nature of monetary policy non-standard measures and finance theory' by Jean-Claude Trichet, 18 November 2010.
        http://www.ecb.int/press/key/date/2010/html/sp101118.en.html
3   See for example (Kahneman, Tversky 1979, Güth et al. 1982, Berg et al. 1995) for interesting studies from the field of behavioural economics.

level, but also as an *in-silicio* test laboratory where we study the outcome of different micro-level specifications.

1.3 Research Question and Outline

The main objective of the dissertation is to contribute to the discussion on the role of demand in the innovation process from an evolutionary perspective. In other words, this dissertation is making a case for the particular role of the demand side, analysing, based on a complex simulation model, the versatile mutual relationships between consumers and producers. As stated before, one cannot expect a full picture of the simulation model. Instead, we stepwise develop an agent-based model of an innovative economy purposely tailored to study in detail the effects emerging if a heterogeneously composed demand side is facing an experimentally organized supply side (Eliasson 1991).

With this, the simulation model carried out in this dissertation follows the key aspects of the evolutionary approach. Instead of oversimplifying the demand side we aim to apply important aspects which too often are only applied to the supply side, e.g. the heterogeneity and bounded rationality of economic actors.

In more detail, the research questions of this dissertation are as follows:

- How are demand side aspects included in economic theory of innovation?
- How can we introduce heterogeneous consumers into a model of an innovative economy?
- What are the implications of considering heterogeneous demand?
- How can network topologies of interactions affect knowledge diffusion and market outcomes?
- What are the effects of considering boundedly rational consumers and which role have consumer networks?
- How do responsible innovations diffuse in markets of heterogeneous consumers?

The analysis which is carried out in the following chapters of this dissertation can be separated in four, closely related and intertwined parts. In the first part, i.e. chapter 2 und chapter 3, we describe the theoretical foundation of the simulation model carried out in the later chapters of this dissertation. We start with a detailed description of how the demand side has been included so far for the analysis of innovation. We hereby explain three different perspectives each of which represented by different models of innovation. In section 2.2 the main characteristics of the evolutionary economics theory as a theoretical framework for the analysis of innovation are discussed. This builds also the theoretical framework of the simulation model elaborated during the later chapters of this dissertation. Finally, section 2.3 outlines the role of the demand side today and discuses some recent approaches and models in the literature which are relevant for this dissertation.

In chapter three, agent-based modelling and its particular role for the scientific endeavour in general and for the model analysed in this dissertation is described. The aim of this chapter is to shed some light on the agent-based modelling approach as a promising and possibly necessary tool in the scientist's toolbox. We focus, therefore, on the most central methodological issues, reviewing the current state of literature and highlighting the possibilities of this still somewhat disputed approach. To acknowledge the novelty of this powerful approach, we start with an introduction to the three pillars of agent-based modelling: modelling, agents and simulation, explaining the common characteristics for all agent-based models. In a next step, important methodological aspects, vital for the successful implementation of an agent-based model are discussed. In the final part of this chapter, the main implications for the later agent-based model are summarised.

After the description of the agent-based modelling approach in chapter 3, chapter 4 describes our first simulation model. With this baseline model, we aim to create a general framework which can display both the heterogeneity of the demand as well as the supply side. We consequently focus first on how the demand of heterogeneous consumers can be incorporated in a computational model. Inspired by the work of Kelvin Lancaster (1966, 1975, 1979), the simulation model incorporates heterogeneity of demand based on heterogeneous consumers equipped with individual preferences for the particularities of product characteristics. With this, we extend the common practice of considering products as characterised only by their price and optionally quality.

Chapter 4 is organised as follows: We start in section 4.1 with some introducing remarks on the scope and the main aim of the agent-based model. In section 4.2 we then describe the basic concepts for products, knowledge and consumer heterogeneity in the model and provide a detailed model description. In section 4.3 we explain and analyse the results of including consumer heterogeneity for the analysis of innovation processes. Our starting point is a standard scenario, which aims to show the fundamental impacts of consumer heterogeneity on innovation dynamics. For this purpose, we focus on the effects of heterogeneity of consumers and vary consumer preferences on a spectrum between full homogeneity and full heterogeneity analysing the behaviour of firms and the resulting market structure. In a second experiment, the simulation model is used as a test laboratory for *in-silicio*-experiments, investigating the impact of different policy strategies designed to foster innovation.

The following chapter 5 introduces a network perspective to the analysis. In more detail, we investigate the importance of network structures in informal firm networks and in consumer networks. To do so, we first implement a barter trade knowledge diffusion process in an agent-based simulation model and analyse how the structural properties of networks affect the overall knowledge diffusion properties within these networks. After some introducing remarks in subsection 5.1.1, subsection 5.1.2 describes in more detail the network topologies used for the analysis and the barter trade diffusion process used. Finally, in subsection 5.1.3 we present the model's results.

In the second part of chapter 5, section 5.2 introduces boundedly rational consumers and, building on this, consumer networks. Introducing heterogeneity of demand presents a huge step towards a more realistic consideration of the consumer into the analysis of innovation processes. However the baseline model in chapter 4 still oversimplifies the capabilities and actions of consumers. After some brief introducing remarks in subsection 5.2.1, we introduce in subsection 5.2.2 consumers with only limited information about firms and products. Consumers in this scenario are not automatically informed about all products and their characteristics but are randomly informed by only a small subset of firms. With this we follow Herbert Simon's idea of boundedly rational economic actors (Simon 1955, 1959, 1972). In subsection 5.2.3 consumer networks are added to the analysis. Based on the findings of chapter 5, we hereby aim to study the role of different consumer network topologies on the innovation processes.

In chapter 6 the simulation model is applied for an analysis of *Responsible Research and Innovation*. As the last extension of our simulation model, chapter 6 implements a second product characteristic which can be considered negative. This negative product characteristic represents the variety of negative effects and aspect of innovation, such as energy consumption, waste, pollution etc. In this scenario, we analyse the effect of more *responsible* consumers (i.e., consumers who consider also negative characteristics) on the innovation process. In more detail, we are interested whether there is an optimal market structure which fosters the production of responsible innovation and, hence, aim to find possible implications to shift markets towards more responsible innovation.

The analysis in chapter 6 follows the structure of the model extensions made in chapter 5. In subsection 6.2.1 we first investigate the models output based on the baseline model from chapter 4. Building on this, subsection 6.2.2 analysis the results in a situation of information scarcity. In the last subsection (6.2.3) consumer networks are added.

Finally, in chapter 7, we discuss a number of important research results as well as possible steps for further avenues of research.

2 The Role of Consumers in Innovation Economics

This chapter describes the theoretical foundation of the simulation model carried out in the later chapters of this dissertation. We aim to show how demand side aspects have been included in the economic literature on innovation in the past and how they are included today. Additionally, this chapter also describes the fundamental aspects of the so called evolutionary economics approach which will be used as a framework for the model carried out in the later chapters of this dissertation.

We start in section 2.1 with a detailed description of how the demand side has been included in the literature on innovation. In more detail, this section discusses three well-known classes or generations of innovation models, each of which, representing fundamentally different perspectives on innovation. In section 2.2 the main characteristics of the evolutionary economics theory are discussed. This discussion also serves as a basic framework for the simulation model elaborated in the later chapter of this dissertation. Finally, section 2.3 outlines today's view on the role of the demand side for innovation, describing related concepts and models from literature.

2.1 The Neglected Demand Side

The following section discusses the relevance of the demand side for the study of innovation by giving insight to three fundamental different perspectives on innovation. While the importance of the demand side has long been acknowledged in various economic fields, its particular role for innovation has long been under great debate. We start in subsection 2.1.1 with some early contributions stressing the concept of the linear innovation model. Based on this, subsection 2.1.2 describes the so called demand-pull model. Finally, subsection 2.1.3 explains the basic ideas behind multidimensional innovation models.

2.1.1 The Linear Innovation Model

As the basis for the later analysis, let us start with a brief look into the history of economists' thoughts on innovation and the demand side. It would be too easy to state that the demand as an important factor is something completely new for the analysis of innovation or in economic science in general. On the contrary, one is tempted to say that the demand side has always been part of the story, the question is in which way it has been considered.

Let us start with a fascinating contribution by Adam Smith who in his famous book: *An Inquiry into the Nature and Causes of the Wealth of Nations* describes the important role of the division of labour for economic growth (Smith 1776).

Although Smith recognizes science and research as a main driver of growth (Knell 2012), he also sees the market size as a crucial because limiting factor:

> "As it is the power of exchanging that gives occasion to the division of labour, so the extent of this division must always be limited by the extent of that power, or, in other words, by the extent of the market." (Smith 1776, p. 31)

He continues and describes for example the particular role of geographical structures and the distribution of population as crucial factors for the division of labour. Smith notices that in the rural and scattered areas of Scotland "every farmer must be a butcher, baker, and brewer, for his own family" in which case there is no place for division of labour (Smith 1776, p. 32). Smith hereby clearly stresses the role of demand as a necessary condition for technological developments which becomes also clear if we look in more detail on the causal loops of economic development.

Following Smith, an increase of market demand fosters the division of labour, sectoral specialisation, and the accumulation of knowledge. This in turn leads to an increased competition between producers and a decrease of prices, at least in the long run. This leads to an increased productivity and revenues which, in turn, increase the market size and demand (Smith 1776, p. 31, see also Knell 2012, Antonelli, Gehringer 2015 for a detailed discussion on this issue). A diagram of this causal loop is shown in Figure 1.

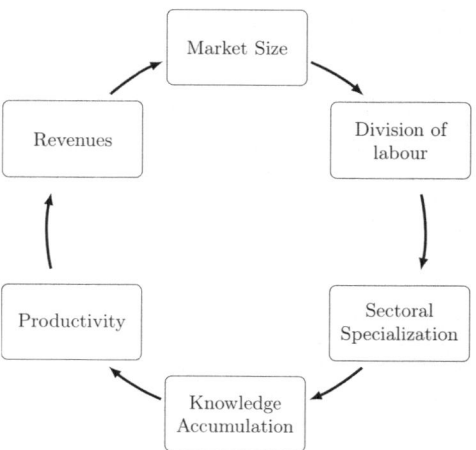

Figure 1:   Smith's Circular Growth (Source: own illustration based on Antonelli, Gehringer 2015).

So, although Smith never particularly stresses the role of demand on innovation, we still see with Smith a general acknowledgment of the size of markets as a driving force for economic development. In a similar vein as the studies by Smith, also other scholars focus their analysis of economic processes on the role of demand. Probably the most famous author in this respect is John Maynard Keynes who focused in his studies deliberately on the role of *demand* which was in contrast to the dominant perspective in economics at this time (see for example Geiger 2015).

In this sense, it is interesting to note that the focus on the demand side has not been adopted by Joseph Alois Schumpeter. Instead, Schumpeter (1912, 1942) focused in his work on the role of monopolists and entrepreneurs stressing that entrepreneurs discovers (new) ideas so far untried and introduces them into the economic realm, leading to *creative destruction* (Coombs et al. 1987). Interestingly, although Schumpeter does acknowledge the influence of consumers' preferences on the production decision of firms, he considers them as static (Knell 2012). The consumer side of the market is driven by routine behaviour and limited foresight, which in turn give the entrepreneurs the role to convince the consumers to change their preferences (Andersen 2007).

From that we can conclude, it would be a false conclusion that Schumpeter refuses any particular role of the demand side. Instead, he as Smith before him recognises the importance of the size of the demand as a factor determining the profitability of innovations. With this, the demand influences the innovative activities of entrepreneurs. However, if asked for the source of innovation Schumpeter clearly would advocate the role of entrepreneurs, large monopolistic or oligopolistic firms.

This passive view on the demand side is also expressed in the famous and well-known linear innovation model. The linear innovation model is something very prominent in economic science and can be traced back directly to the ideas and the work of Schumpeter (Godin 2006). As a matter of fact, it is highly doubtful that there is a student who has not seen it, or can avoid it in his studies. It is a model of a theory, demonstrating and underlining that innovation is indeed a process with several definable steps which are embedded in a multifaceted innovation process (Godin 2006).

Early types of linear innovation model, can be traced back to the work of Price and Bass (1969) or Langrish et al. (1972), others say that the origins go even back to Bush (1945) *Science: The endless frontier* (see also Godin 2006 for a great overview on the history of the linear innovation model). In the literature we find a wide range of different types and forms of the linear innovation model in the literature. In principle, we can identify in linear innovation models three separate building blocks: the nature of the sources of innovation, the innovative process, and the effect of innovation (Kline, Rosenberg 1986, Rothwell 1994, Edgerton 2004, Godin 2006).

In its original form, the linear innovation model follows the innovation process as introduced by Schumpeter. A common composition of such a model can be seen in Figure 2. In this case we differ between 5 steps which run sequentially, i.e. basic research, applied research and development and ends with production and diffusion (see for example Godin 2006). In this model, the two elements basic research and applied research act as a source for innovation. The innovation process then is described via the development and production. Finally, the effect of innovation is seen in the diffusion of the innovation.

Figure 2:   The Linear Innovation Model (Source: own illustration based on Godin 2006).

The exact composition of the elements of the linear innovation has long been under great debate. For example, it is questionable to what extend we can make a clear cut between basic and applied research. The key characteristic of this model, however, remains. It is a stepwise process, going sequentially from one end to the other, and hereby putting special emphasis on research and development as the source of innovation reflecting Schumpeter's perspective on innovation.[4] The demand side has no particular role from this perspective. At the best, this early perspective considers the demand side as something passive and pale.

2.1.2   The Demand-Pull Modell of Innovation

Interestingly, it was not until the seminal work of Jacob Schmookler (1962, 1966) and Nicholas Kaldor (1966, 1972) that the demand side has been (re-) recognised as one of the key factors of innovation and economic development. Although Schmookler and his studies are generally credited to be the reason for the new momentum in the debate on the role of the demand side, there is also the work of Nicholas Kaldor (1966, 1972), who contributed important insights in a post-Keynesian framework analysis. So for example in (1975, p. 895) Kaldor stated:

> "[…] economic growth is demand-induced, and not resource-constrained - i.e. that it is to be explained by the growth of demand which is exogenous to the industrial sector' and not by the (exogenously given) growth rates of the factors of production, labour and capital, combined with some (exogenously given) technical progress over time."

---

4   This is also the reason why Schumpeter often is considered as the intellectual father of this model (Godin, Lane 2013).

Kaldor emphasises in his papers multipliers and accelerator effects to support the idea that increases in public expenditures, able to support the expansion of aggregate demand, have positive effects on productivity growth, output and eventually investments (Antonelli, Gehringer 2012). Additional investments, in turn, were expected to fasten the diffusion of technological innovations. In a similar vein as in Figure 1, Figure 3 presents the Kaldorian demand-pull mechanism as a circular process of cumulative causation:

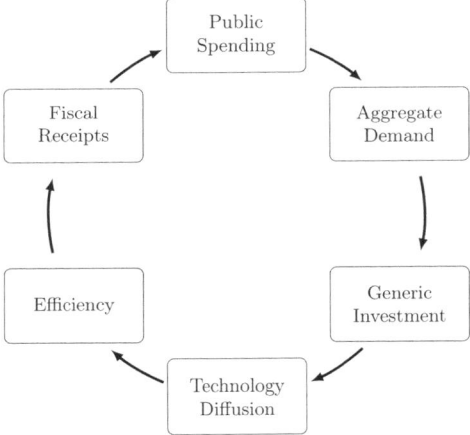

Figure 3:   The Kaldorian Demand-Pull Mechanism (Source: own illustration based on Antonelli, Gehringer 2012).

Unfortunately, the model of economic growth by Kaldor did not receive much attention at its time. Instead, it was Schmookler, who gave new momentum to the story by contributing one of the first empirical studies on the relationship of demand and innovation. Unsatisfied with the linear view of a supply-oriented focus on the innovation process, Schmookler (1962, 1966) analyses empirically the connection between demand and innovation studying different industries of the US economy.

In his paper *Economic Sources of Inventive Activity* (1962) Schookler uses patent data to analyse the relationship between demand and innovation[5] in the railroad industry. His results show that peaks in patenting activities indeed lag behind peaks of production of commodities, from which he concluded that innovation can no longer be regarded as independent variable as it was done so far. He concluded that the influence (upon innovation) of the latter (unfolding

---

5   Technically speaking, Schmookler, thus, does not directly measure the effect on innovation but on inventions.

economic needs) has been substantial, at least in established industries (Schmookler 1962, p. 20). Schmookler's explanation for this fact is simple, yet, convincing:

> "[…] the incentive to make an invention […] is affected by the excess of expected returns over costs. Scientific progress may reduce expected costs and so increase the probability that a given invention will be sought and made. However, every invention represents a fixed cost, and the expected benefits from it vary with circumstances." (Schmookler 1962, p. 19)

In Schmookler's perspective, the size of the potential market represents a necessary trigger for firms to innovate. The potential market determines the potential benefits of an innovation and directly influences the innovative behaviour of profit seeking firms. In other words, it would be highly doubtful that firms engage in any costly innovative activity without the expectation to gain profits from that innovation. In fact, consumers are the driving force for any innovation activity, steering firms towards what they want.

Schmooklers studies, but also famous studies such as *Project Hindsight* by the United States Department of Defense (Sherwin, Isenson 1967) and many others (see for example Mowery, Rosenberg 1979, Rothwell 1977, Coombs et al. 1987 for an overview) led to a second class of sequential innovation models, i.e. demand-pull models which are strongly related to previous linear innovation models.[6] Early versions of a demand-pull model emerged in literature in the late 1970s. They reflect the upcoming debate in the 1960s on the influence of the demand side and are actuated by the findings of Schmookler (Godin, Lane 2013). For example in (Rothwell, Zegveld 1985) we find a diagram of the demand pull model as pictured in Figure 4. In this case the source of innovation clearly lies in market needs. The development and manufacturing, instead, can be seen as the process of innovation. Finally, the element sales represent the effect of an innovation.

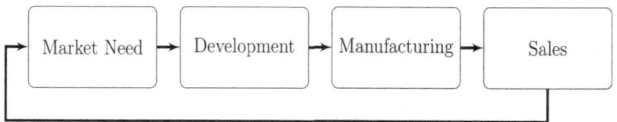

Figure 4:   The Demand Pull Model (Source: own illustration based on Rothwell, Zegveld 1985).

Although the demand-pull model shares the sequential perspective with the former linear innovation model, it still represents a major shift in the underlying perception of the innovation process. It puts emphasis on the market needs as the

---

6   The demand-pull model sometimes also is labeled as *Market-Pull* or *Need-Pull* model (Rothwell 1992).

nature of the sources of innovation instead of basic science or research and development as in the previous form. Here it is argued that most critical for innovations are need-pull forces expressed in opportunities, pulling from peoples' needs (Godin, Lane 2013). In other words, following this perspective the demand side is something active which represents a huge change in thinking for many scholars from that time.[7] Unfortunately, the demand-pull model was only for a short period in the focus of scientific literature and has been replaced, together with the technology-push model, by the upcoming multidimensional innovation models in the 1980th (Rothwell 1992, Godin, Lane 2013).

The main reason for this change was the dissatisfaction with the sequential nature of both models. The critique that innovation is not a linear process, however, was nothing new. So did for example already Price and Bass (1969, p. 802-803) note:

> "[…] innovation is often viewed as an orderly process, starting with the discovery of new knowledge, moving through various stages of development, and eventually emerging in final, viable form. According to this 'linear' model, innovation seems to be a rational process […]."

Analysing and reviewing three studies on innovation Price and Bass yet come to the conclusion:

> "The studies reviewed here indicate that the 'linear' model is not typical. One appreciates the nonrational nature of the innovative process when one notes that the more novel the invention is, the less orderly and predictable is the process." (Price, Bass 1969, p. 803)

Additional points of critiques on the existing models of innovation were for example the distinction between basic (scientific) and applied (technological) research, and how the former informs the latter and the connection between transfer of new scientific knowledge into technology and commercial innovations (Balconi et al. 2010).

In summary, the demand-pull model introduced a new perspective to the question about the main drivers of innovation (see Coombs et al. 1987 for a good overview about the four central question about the origin of innovations). The vast range of critics led to an intense debate which was calmed by the idea that the question for an ultimate source of innovation does not lead to any further insights. Instead, scholars began to understand that innovation has a strong systemic character (see for example Fagerberg 2004) in which multiple elements are in

---

7   To put it to an extreme, the demand side in the demand-pull model still represents something quantitative: more demand means more innovation. Admittedly, a more qualitative view on the causal interdependencies between demand and innovation is also in the demand-pull model missing.

strong interdependence. This understanding quickly led to the development of the third generation of innovation models: multidimensional models of innovation.

2.1.3 A Multidimensional Perspective on Innovation Processes

Since the seminal work of Schmookler, a series of empirical studies have tried to support the hypothesis by Schmookler (e.g. Sherwin, Isenson 1967, Myers, Marquis 1969, Langrish et al. 1972, Gibbons, Johnston 1974, Gibbons, Gummett 1977, Scherer 1982, see also Andersen 2007) while others refused this perspective and viewed the activities and internal capabilities of firms as the primary drivers of innovation. The two most common critiques in this respect have been that (1) Schmookler deals with invention rather than innovation and (2) that demand-pull has often been interpreted in terms of need-pull which is a concept too broad to allow for falsification (see for example Mowery, Rosenberg 1979, Walsh 1984, Kleinknecht, Verspagen 1990).

As a preliminary result of this debate between proponents of demand-side arguments and their opponents, Mowery and Rosenberg (1979) state that this debate is misleading and that both the demand and the supply side appear to simultaneously play crucial roles:

> "Rather than viewing either the existence of a market demand or the existence of a technological opportunity as each representing a sufficient condition for innovation to occur, one should consider them each as necessary, but not sufficient, for innovation to result; both must exist simultaneously."

Following Mowery and Rosenberg successful innovation emerge from the fruitful interaction between demand-pull and technology-push effects and consequently no dominant factor can be identified (Mowery, Rosenberg 1979).[8] In other words, while without doubt the demand side plays a major role in shaping the direction of scientific progress, it does so within the changing limits and constraints of a body of scientific knowledge (Rosenberg 1974). In a similar vein, also Giovanni Dosi (1982, 1988a) sees the demand side as in important factor within the concept of technological paradigms:

> "The evidence on market-induced innovative activity [...] may indeed be consistent with our model [...]. This process, however, relates much more to normal technology than to discontinuous technological advances." (Dosi 1982, p. 159)

---

8   Admittedly, although the appeasing simultaneous recognition of supply-side and demand-side is generally attributed to Mowery and Rosenberg (1979), already Schmookler (1966, p. 11) considered that: "Without wants no problem would exist. Without knowledge they could not be solved".

This shift in the understanding that innovation is not an orderly stepwise and sequential process with a beginning and an end led to the development of so called multidimensional models.[9] As an example, we briefly discuss the structure of the famous chain-linked model by Kline and Rosenberg (Kline 1985, Kline, Rosenberg 1986).

The chain-link model is one of the first representative examples of the class of multidimensional models which are characterised by interactions and feedback loops among all the factors involved in the process of innovation. Multidimensional models, therefore, avoid the question for an ultimate source of innovation. As also Kline (1985, p. 44) puts it:

> "In looped processes, every cause becomes in due time an effect, and every effect becomes in due time a cause. The distinction between pushes and pull loses all meaning."

The key issue here is that in contrast to the two linear models discussed so far, the chain-link model by Kline and Rosenberg manages to combine the mutual existence of demand-pull and technology-push effects in the innovation process. With this a clear shift in the debate about the role of the demand side becomes evident, while previous research focused mainly on identifying the main causes for innovation. The work of for example Mowery and Rosenberg (1979), Kline (1985), Kline and Rosenberg (1986) redirected the debate to the qualitative interdependence between all factors. This opened the discussion for questions not *if* demand is important for innovation, but *how* it is and how innovation is important for demand, stressing the systemic character of the innovation process (see also Fagerberg 2004).

---

9   Multidimensional models also go under the terms iterative, interactive, recursive, systemic etc. Some important examples are for example the *Multidirectional model* (Pinch, Bijker 1987), *Neural net model* (Ziman 1991), *Coupling model* (Rothwell 1992), *Interactive model* (Newby 1992) and *Linear-plus model* (Tait, Williams 1999). See also Godin and Lane (2013) on this issue.

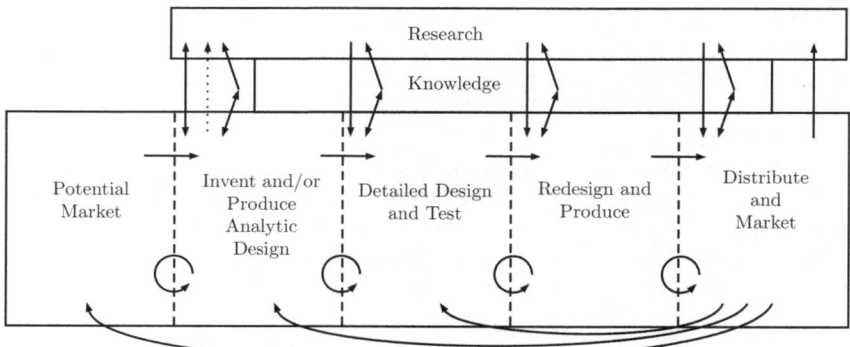

Figure 5: The Chain-Link Model (Source: own illustration based on Kline, Rosenberg 1986).

The chain-linked model by Kline and Rosenberg (see Figure 5) consist of seven elements embedded in a system without a clear and defined order:

- Research
- Knowledge
- Potential market
- Invent and/or produce analytic design
- Detailed design and test
- Redesign and produce
- Distribute and market

In their model, Kline and Rosenberg consider several feedback loops and couplings between the elements. Additionally, they identify not one but five major paths of activity (see Kline, Rosenberg 1986 for a detailed discussion about each path). So for example one path consists of starting with the potential market, ending up with the distribution. However, other paths are possible and can have different sources and elements of the system. For all of them, Kline and Rosenberg considered the possibility of several feedbacks and loops which are indicated by the various arrows in Figure 5. These feedbacks and loops illustrate that no clear sequential order is in action and correspondingly there are no such things as the source or the effect of innovation.

Kline and Rosenberg (1986, p. 291f) describe the role of the demand side in the chain-linked as follows:

> "A perceived market need will be filled only if the technical problems can be solved, and a perceived performance gain will be put into use only if there is a realizable market use. Arguments about the importance of 'market pull' versus 'technology push' are in this sense artificial, since each market need entering the innovation cycle

> leads in time to a new design, and every successful new design, in time, leads to new market conditions."

In other words, in the multi-dimensional model by Kline and Rosenberg the demand side and the supply side are in a coevolving relationship. Although some authors may criticise this ambiguity[10], the coevolution of demand and supply today is widely excepted, yet, underexplored (see for example Saviotti, Pyka 2012). Well-known approaches to get a in depth understanding on the relationship are for example the concept of democratizing innovation by Eric von Hippel (1976, 1988) or the wide field of studies about the diffusion of innovation which will be explained in more detail in section 2.3. However, by far the larger portion of work on innovation and technological change is concentrated on supply-side dynamics (see for example Adner, Levinthal 2001, Coombs 2001, Witt 2001a, Harvey et al. 2001, Andersen 2007, Ciarli et al. 2008, Nelson, Consoli 2010).

Summing up this section's findings, the way the literature considers the demand side and its effect on innovation considerably changed over time and was always driven by a fruitful discussion. The debate led to an understanding based on which the demand and the supply side are in a mutual relationship with different causal processes. The analysis of this dynamic relationship, however, requires both: a theoretical and a modelling framework which allows us to systematically analyse the mutual interaction between two too often isolated elements of the complex economic system.

2.2 An Evolutionary Perspective

The term evolutionary and its connection to economy theory has long been under great discussion and unfortunately has not achieved at a final consensus yet. For the beginning, let us start with a broad understanding from Boulding (1991, p. 9) who says that: "evolutionary economics is simply an attempt to look at an economic system [...] as a continuing process in space and time". Although this definition may convince through its simplicity and parsimony, there is more to the evolutionary approach which is relevant and makes the evolutionary perspective the right approach to analyse innovation processes.

Instead of a broad agreement, today, the debate on what evolutionary economics has spread into several different groups, encompassing a variety of perspectives (Witt 2008, Hodgson 1999). In fact, the use of evolutionary as a term in economics is so wide-spread that for example Hodgson understands this as "a

---

10   For example Godin and Lane (2013, p. 24f), criticise that Kline and Rosenberg reduce the meaning of the demand side to what it was before the demand-pull model: "a single factor (among many) [...]".

matter of fashion" (Hodgson 1999, p. 128). Also Witt (2008) distinguishes based on the ontological stance and the heuristic strategy between the four distinct research strands: Universal Darwinism, Naturalistic, Schumpeterian and neo-Schumpeterian approaches which deserve to be labelled as evolutionary. This of course makes an exact definition of what evolutionary economics is and what the most important aspect in general are beyond the reach of this dissertation (see also Witt 2008 for an interesting questionnaire). However, we can capture three important elements of all evolutionary strands: an evolutionary theory is (1) dynamic, (2) historical and (3) self-transformation explaining (Witt 1993). In the following, we will briefly describe major cornerstones on the way towards an evolutionary economic theory, and thus explaining some of the main characteristics of the methodological framework of the model analysed in this dissertation.

Although the discussion about evolutionary economics gained momentum only since the second half of the twentieth century, the roots of an evolutionary perspective thinking can be traced back in economic histories to the late 19th century and is found even amongst authors who are well known for their contribution to equilibrium theories (Dosi, Nelson 1994).

One of the first authors to deliberately mention evolutionary ideas in his work is Thorstein Veblen who already in 1898 demands in his paper "*Why is Economics not an Evolutionary Science*" an evolutionary economic theory (Veblen 1898, see also Hodgson 1998b for a detailed discussion on Veblen´s contribution). Highlighting especially the process characteristic of evolutionary economics, Veblen (1898, p. 375) frames his understanding of an evolutionary economics theory as follows: "Any evolutionary science […] is a close- knit body of theory. It is a theory of a process, of an unfolding sequence". And at a later part of his paper, he continues:

> "it appears that an evolutionary economics must be the theory of a process of cultural growth as determined by the economic interest, a theory of a cumulative sequence of economic institutions stated in terms of the process itself." (Veblen 1898)

Interestingly, already Veblen (1898) who is usually associated with other strands of economic theory, stresses the process character of an evolutionary economics theory. Veblen, however, has not been the only early contributor to the discussion. Evolutionary ideas can also be found at well-known economists such as Smith, Malthus, Marx, Penrose, Alchain and Marshall (see also Nelson, Winter 1982, Dosi, Nelson 1994, Hodgson 1999, Potts 2003).

One of the cornerstones for the development of an evolutionary theory of economics, and as some say maybe the biggest one (Fagerberg 2003, p. 11), can be credited to Joseph Alois Schumpeter and his work (e.g. Schumpeter 1912,

Schumpeter 1942). Although Schumpeter avoided the term *Evolutionary* or *Evolution* in his early work (Witt 2008, p. 554), his unique interpretation of the economic development can only be considered a path-breaking contribution to evolutionary economics, which made him for almost 50 years the leading academic protagonist for the evolutionary approach (Fagerberg 2003).

In his famous book from 1912 *Theory of Economic Development* Schumpeter challenged the dominant view at this time that the cause for economic development can be found in exogenous factors. Instead, Schumpeter focused on endogenous drivers of economic development, providing the theoretical basis for a new paradigm (see for example Hanusch, Pyka 2007b for a discussion on Schumpeter´s contributions). Especially in his early work, Schumpeter highlights the role of the entrepreneurs as driving forces for evolutionary change, introducing the concept of the *creative destruction* as an inherent force that destroys every state of equilibrium (Kwasnicki 2007, Hanusch, Pyka 2007b). With this, Schumpeter as did Veblen beforehand, directly addresses especially the dynamic character of economic processes (Witt 2002), highlighting and emphasizing the role of innovation for economic processes. However, Schumpeter differs from Veblen in a sense that Schumpeter takes a perspective which Witt calls a dualistic ontology, treating economic and biological evolutionary processes as belonging to different, disconnected, spheres of reality (Witt 2008).

The next leap in the development of the evolutionary framework in economics can be seen in the work of Nelson and Winter and their synthesis of evolutionary ideas framing a comprehensive evolutionary approach in economics. A central motivation for Nelson and Winter's work rests on the limiting and unrealistic assumptions of what Nelson and Winter call traditional models (e.g. Nelson, Winter 1973) or, in more general, orthodox economic theory of their time (e.g. Nelson, Winter 1982). The notion of Orthodoxy as used in later work of Nelson and Winter widely refers to "a modern formalization and interpretation of the broader tradition of Western economic thought whose line of intellectual descent can be traced from Smith and Ricardo through Mill, Marshall, and Walras" (Nelson, Winter 1982, p. 6).

Nelson and Winter can be credited for at least three important papers to the discussion in the 1970s (Nelson, Winter 1973, Nelson, Winter 1974, Nelson, Winter 1975), which also constitute the underpinnings of their famous book *An Evolutionary Theory of Economic Change* from 1982. Inspired by the existence of an expanding literature that abandoned the at this time dominating models and their structural pillars and in favour of more sensitive and subtle formulations, Nelson and Winter criticized the restrictive and limiting setting of the orthodox framework (Nelson, Winter 1973). In particular, the authors rejected that the orthodox framework builds on the general assumption of maximisation as the key driver for firm behaviour within an analytical framework based on general economic equilibria (Nelson, Winter 1982).

Although Nelson and Winter (1982) base their description of an evolutionary economics theory also on an operational concept of the decision rules employed by firms, they strongly reject the idea of maximisation as the main driver for firms' behaviour. Instead, Nelson and Winter highlight the concept of so called routines to describe firms' activities. Routines in the sense of Nelson and Winter (1982, p. 14) cover:

> "[...] characteristics of firms that range from well-specified technical routines for producing things, through procedures for hiring and firing, ordering new inventory, or stepping up production of items in high demand, to policies regarding investment, research and development (R&D), or advertising, and business strategies about product diversification and overseas investment."

It is important to note that the contributions of Nelson and Winter cannot be seen isolated from the work of Schumpeter. In contrast, the connection of Schumpeter's work and the work by Nelson and Winter goes so far that the evolutionary approach carried out by Nelson and Winter can be seen a neo-Schumpeterian:

> "The influence of Joseph Schumpeter is so pervasive in our work that it requires particular mention here. Indeed, the term 'neo-Schumpeterian' would be as appropriate a designation for our entire approach as 'evolutionary' " (Nelson, Winter 1982, p. 39).

Nonetheless, we would go too far stating that the evolutionary economics approach in general is the same as neo-Schumpeterian approach. Given the wide range of different roots of evolutionary economics and thus, the different foci we agree with Hanusch and Pyka (2007a) and consider evolutionary economics as the basic framework in which neo-Schumpeterian economics puts a special focus on innovation.

In fact, there are some important differences between the original contribution by Schumpeter and the work of Nelson and Winter (see also Hodgson 1997, Witt 2002, Kwasnicki 2007 for a detailed discussion).[11] Nelson and Winter, in contrast to Schumpeter, include the general ideas of natural selection using a heuristic that makes metaphorical use of Darwinian concepts (Dosi, Nelson 1994, Fagerberg 2003, Witt 2008). In their famous book, Nelson and Winter (Nelson, Winter 1982, p. 9) promote the idea of general selection as the main characteristic of their approach:

---

11   Some authors differ in this context sometime between *the old evolutionary economics* and *the new wave of evolutionary theorists* (see for example Hodgson 1993, Andersen 2013, Fagerberg 2003).

> "Our use of the term 'evolutionary theory' to describe our alternative to orthodoxy
> [...] is above all a signal that we have borrowed basic ideas from biology, thus
> exercising an option to which economists are entitled in perpetuity by virtue of the
> stimulus our predecessor Malthus provided to Darwin's thinking."

As Nelson and Winter note, evolutionary economics takes a process perspective on economic systems which stands in contrast to the equilibrium assumption of orthodox economic thinking[12]. Evolutionary economics in this context focuses on the explanation of the movement of an economic system, (as a whole or only parts of it) over time instead of analysing the dynamics of static equilibria (Dosi, Nelson 1994, Nelson, Winter 1982). This dynamic perspective of evolutionary economics, allows to consider path dependencies and the irreversibility of economic processes (David 1985, Arthur 1989, Hanusch, Pyka 2007a). The framework of the evolutionary economics approach enables us to study the versatile dynamics of the coevolution of the demand and the supply side.

The evolutionary economics approach also includes a sophisticated and more accurate description of economic actors and their behaviour. As for example Nelson and Consoli (2010, p. 666) note:

> "[evolutionary economists] share not only an interest in building a theory that deals
> with economic dynamics better than does neoclassical theory. They also share a set of
> beliefs in how to characterize and understand human behavior, and the behavior of
> human organizations, that leads them to reject neoclassical theory not only as a
> framework inadequate for understanding economic dynamics, but more generally as
> a deeply flawed theory of economic behavior in any context [...]."

If we want to understand the mutual relationship between demand and supply, we have to consider demand side and supply side as more than just aggregated forces within the economic system. Instead, we focus on the entities behind these forces, e.g. consumers and firms which are of great relevance for a thorough and comprehensive understanding. Let us start with an aspect of evolutionary economics which is also considered at the later chapters and especially for the simulation model, namely the concept of heterogeneity. The evolutionary economics approach as for example described by Nelson and Winter breaks with the tradition and common practice to extrapolate the characteristics of a representative agent to an entire population (often referred to as *typological thinking*). In contrast, Nelson and Winter's *population thinking* looks at the social and economic consequences of interaction within population s of heterogeneous actors (Mayr 1959, Hodgson 1993, Andersen 2013).

---

12   In this context, evolutionary economist often refer to so called *punctuated equilibria* (Hodgson
      1997).

The aspect of heterogeneity is one of the most import aspects of evolutionary economics (Cantner, Hanusch 1999) and is in full contrast to the neoclassical approach and its concept of representative agents or representative firms (Marshall 1890). The concept of heterogeneity in evolutionary economics covers aspects such as: behaviour, attitudes or characteristics of agents (Cantner, Hanusch 2001). In its basic essence acknowledging the heterogeneity means recognizing economic actors as what they are: individuals facing *Knightian true uncertainty* (Knight 1921, Nelson, Winter 1982, Dosi, Nelson 1994).[13]

Building on this, Nelson and Winter include the concept of *procedural* or *bounded rationality* as described by Herbert Simon (1957, 1972) into their concept of evolutionary economics as a more elaborate theoretical perspective on how firms behave (Fagerberg 2004). Instead of perfectly rational behaving entities, firms in the perspective of Nelson and Winter behave guided by *routines* as heuristics and simple rules of thumb to tackle decisions. To evaluate the performance of these heuristics, firms apply a so called satisficing behaviour. So if a routine leads to an unsatisfactory outcome, e.g. the profits or sales fall below a certain threshold firms search for a new routines, which will eventually be adopted if it satisfies the criteria set by the firm (Fagerberg 2004).

As a result of the previous discussion, it is important to note that the perspective on evolutionary economics as proposed by Nelson and Winter, and which is today often seen as the main driver for the new momentum evolutionary economics gained today, cannot be considered as an independent and self-contained approach. Instead, it is the synthesis of different evolutionary scholars, and amongst them especially the work of Schumpeter and the growing demand for alternative approaches getting loose from the restrictive frameworks of orthodox models. As we have seen, the evolutionary economics approach offers a unique perspective both on a systemic level as also on an actor level. On a systemic level it breaks with the limiting equilibrium framework used by other concepts and capture a dynamic, process oriented view. As for example Hanusch and Pyka (2007a, p. 280) point out:

> "The outcome of evolutionary processes is determined neither ex-ante nor as the result of global optimizing, but rather is due to true uncertainty underlying all processes of novelty generation, and so allows for openness towards future developments - a feature of evolutionary theories which makes them ideal for analyzing innovation processes."

---

13   In contrast to situation of risks, true uncertainty refers to situations where not only the consequences of the set of possible alternatives are indefinable but it is also unknown which alternatives exist (Keynes 1937, Nelson, Winter 1982).

Second, evolutionary economics get rid of the oversimplified assumptions about economic actors which become necessary in equilibrium frameworks. Instead, within evolutionary economics it is possible to treat economic actors as individuals which are heterogeneous and only boundedly rational. This unique perspective offers a framework to analyse the mutual relationship between something heterogeneous and dynamic as innovation and demand.

2.3 The Role of the Demand Side Today

The debate of the last decades between the demand-pull and supply-push arguments has today not completely vanished. However, it is still true as many authors argue that demand-side effects on the innovation processes have been somewhat neglected or disregarded over the last decades (Coombs 2001, Witt 2001b). More specifically, Adner and Levinthal (2001) note that by far the larger portion of work on technological change is concentrated on supply-side dynamics. Harvey et al. (2001) stress the myopic concentration on the terms of market exchange characteristics of innovation studies, with its excessive attention to supply-side processes. The debate today has branched out into a number of sub-debates, each highlighting different aspects of demand in the innovation processes. With this, we see a more comprehensive analysis of the multifaceted relation between demand and innovation as for example Anderson (2007) points out. The following section discusses in some more detail the focus and contributions of some different branches.

Maybe the most prominent example in this respect has been the study of the diffusion of innovations. Early contributions to this field of study include the work of e.g. Gabriel Tarde and Friedrich Ratzel and Leo Frobenius in the late 19th century to the beginning of the 20th century. The empirical groundwork for the theory of innovation diffusion, however, was laid by Ryan and Gross (1943), who found that social contacts, social interaction, and interpersonal communication were important influences on the adoption of new behaviours (Valente, Rogers 1995). Today's interest in this field of study can be traced back in particular to the book *Diffusion of Innovation* first published in 1962 (Rogers 2010). For Rogers the diffusion of innovation means: "the process by which an innovation is communicated through certain channels over time". Rogers hereby refers to a fundamental aspect of innovation which already Schumpeter stressed (Schumpeter 1928, p. 378):

> "What matters […] is merely the essentially discontinuous character of this process, which does not lend itself to description in terms of a theory of equilibrium". In other words, innovations do not automatically diffuse through economic systems. Instead,

we have to consider that some innovations diffuse only slowly and some not at all. So the question emerges how and why do innovations diffuse?

Common elements of this interdisciplinary field of study and possible aspects to tackle these questions are: the innovation itself, the population of potential adopters and innovators and the flow of information about the innovation between manufacturers and adopters (Coombs et al. 1987).

One of the central assumptions of this framework is that adopters pass through a process of five stages: knowledge, persuasion, decision, implementation and confirmation which form the adoption decision of consumers (Rogers 2010). Building on this, it is possible to specify a large number of different factors determining the diffusion of innovations. These include innovator characteristics, particular innovation characteristics, characteristics of consumers and the characteristics of the information network between consumers and firms (see also Wejnert 2002).

One of the basic concepts within the field of diffusion research is the categorization of adopters in different adopter categories (Rogers 2010). In Figure 6 we show the resulting distribution measured by the time at which an individual adopts an innovation. Based on the propensity to adopt we can distinguish between five different adopter categories: innovators, early adopters, early majority, late majority and finally laggards. The boundaries of the categorization are generally defined by the average time an innovation is adopted ($\bar{x}$) and the standard deviation (sd).

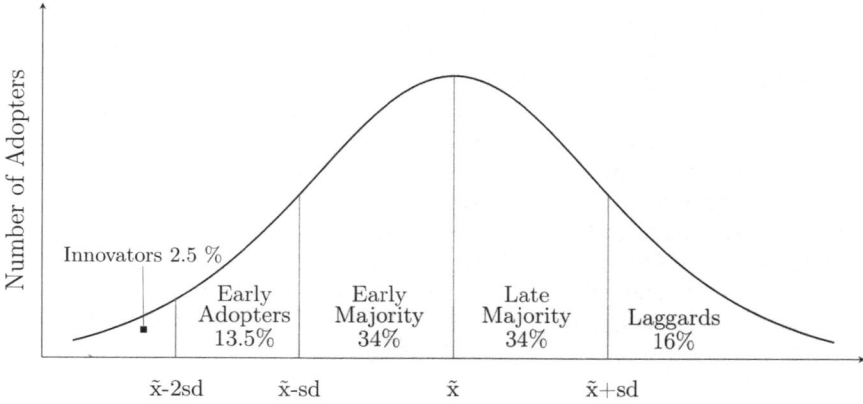

Figure 6:   Adopter Categorization after Rogers (Source: own illustration based on Rogers 2010).

The studies in the field of innovation diffusion without doubt gave new interesting insights to the particular role of consumers (or to be more precise adopters).

However, it has often been criticised that potential adopters are too homogeneous in the analysis and the system itself behaves statically (Coombs et al. 1987, Kiesling et al. 2012). So for example, especially aggregated models (e.g. the famous diffusion model by Frank Bass (1969)) of the diffusion of innovations are based on the assumption that adopters are held homogeneous.

Today various individual-based diffusion models have extended the scope and consider the heterogeneity of adopters. For diffusion models, the heterogeneity of adopters is mainly modelled via heterogeneously distributed adoption thresholds to approximate varying propensities to adopt a new technology (Kiesling et al. 2012). Other important approaches include heterogeneity of demand through reservation prices (see for example Cantono, Silverberg 2009), communication behaviour (Rahmandad, Sterman 2008) or sociodemographic characteristics (see for example Dugundji, Gulyás 2008).

Insights from studies on the diffusion of innovation are also the basis for the famous lead-user approach by Eric von Hippel (1986, 1988). Combined with concept of so called user-innovation both aspects are generally summarised by the term *democratizing innovation*. Taking into account that consumers are heterogeneous in terms of their needs von Hippel describes the chance and importance of identifying so called *lead-users* who face needs months or years before the bulk of a marketplace encounters these needs and expect to benefit considerably by the new innovation. Although this concept is often used in marketing and management science it represents a without doubt important example for the shift in economists' minds. First, recognising consumers as important sources for information stands in contrast to the assumed passiveness of consumers underlying many economic approaches. Second, the lead-user concept also acknowledges heterogeneity of consumers, giving proof that the generalising approach of representative consumers neglects important aspects of reality.

Additionally, Eric von Hippel's user-innovation approach shows that the strict separation between consumers, i.e. economic actors purchasing innovations and firms as economic actors producing innovation not always holds true. Several studies show that consumers often take an active role in the innovation process and also become inventors themselves. Important examples are mountain bikes, several open source software products, the Go-Pro camera, Q-tips and many more. Unfortunately, this special form of invention and innovation got only little attention by the broad majority of innovation researchers so far (von Hippel 1988).

Another strand of important literature which gave great inspiration to this dissertation are agent-based models of innovation. In recent years a small number of models have extended the scope building on an evolutionary economics framework and analysing the relevance of different demand side aspects and the effects on their innovative process (see for example Andersen 2001, Saviotti 2001,

Metcalfe 2001, Saviotti, Pyka 2004, Ahrweiler et al. 2004, Valente 2012, Babutsidze 2012, Lorentz et al. 2015 for important contributions). The following brief summary discusses some of them in more detail.

One interesting example for a model of innovation in a neo-Schumpeterian fashion is the so called SKIN (*Simulating Knowledge in Innovation Networks*) model family by Nigel Gilbert, Petra Ahrweiler and Andreas Pyka (shown for example in Gilbert et al. 2001, Ahrweiler et al. 2004, Gilbert et al. 2007). SKIN is an agent-based simulation platform in which a special focus is placed on knowledge and learning of firms. The model considers different strategies of learning of firms, i.e. incremental and radical, learning-by-doing but also networking.

The model draws on the concept of *kenes* (Gilbert 1997) to represent the individual knowledge stock of a firm. A firm j's knowledge stock $\overline{K}_j$ is represented as a set of knowledge units $K_{i,j}$ depicted as triples of the capabilities $C_{1,...,n}$, abilities $A_{1,...,n}$ and expertise levels $E_{1,...,n}$ of that knowledge unit (see also equation 1).

$$\overline{K}_j = \begin{pmatrix} C_1 \\ A_1 \\ E_1 \end{pmatrix}, \begin{pmatrix} C_2 \\ A_2 \\ E_2 \end{pmatrix}, \begin{pmatrix} C_3 \\ A_3 \\ E_3 \end{pmatrix}, ..., \begin{pmatrix} C_n \\ A_n \\ E_n \end{pmatrix} \tag{1}$$

In this context, the respective capability refers to the general technological or business domain (e.g. biochemistry). Its ability is the indicator for the application in this field (e.g. a synthesis procedure or filtering technique in the field of biochemistry). Finally, the expertise level gives information about the expertise of the firm in that particular field gained so far (Ahrweiler et al. 2004). To give a more detailed example we show in equation (2) an exemplary knowledge stock $\overline{K}_j$ of firm j:

$$\overline{K}_j = \begin{pmatrix} 1 \\ 4 \\ 2 \end{pmatrix}, \begin{pmatrix} \mathbf{47} \\ \mathbf{8} \\ \mathbf{9} \end{pmatrix}, \begin{pmatrix} 56 \\ 2 \\ 7 \end{pmatrix}, \begin{pmatrix} \mathbf{143} \\ \mathbf{9} \\ \mathbf{1} \end{pmatrix} \tag{2}$$

Furthermore, it is assumed that the knowledge owned by a firm will not be entirely used for the production of a certain good. There are knowledge pieces $IH_{i,j}$ in the knowledge base of a firm which are well known to the firm, but are at a particular moment in time unnecessary for production. Knowledge pieces which are used for the production of goods are part of a so called *innovation-hypothesis* (Gilbert et al. 2001, Ahrweiler et al. 2004).

$$IH_j = \begin{pmatrix} 47 \\ 8 \\ 9 \end{pmatrix}, \begin{pmatrix} 143 \\ 9 \\ 1 \end{pmatrix} \qquad (3)$$

Based on the information given in the innovation hypothesis the model computes the product category, its quality and its price. A particular feature of the SKIN model is its endogenous representation of demand. In addition to end consumer demand, each firm in the model demands input products in order to be able to produce and sell products which are also computed based on the firm's innovation hypothesis. These input products, in turn, are products which need to be produced by other firms. As a result, the model creates highly dynamic value-chains in which demand and supply are coevolving over time. However, despite the interesting dynamics created, products in the SKIN model are only characterised by their individual product category, their prize and the quality which leaves only limited possibilities to consider for example demand heterogeneity.

Another interesting approach, related to the model elaborated in the later chapters, is the model by Marco Valente which has been used in several papers (see for example Valente 1999, Valente 2009, Bleda, Valente 2009, Ciarli et al. 2010, Valente 2012). Products in the model by Valente are represented as vectors over a set of dimensions representing different product characteristics. In Table 1 the generic value v_X^i is the measure of product X in respect of characteristic i.

Table 1:   Products' Quality Values (Source: own illustration based on (Valente 2009).

	Char. 1	Char. 2	...	Char. m
Prod. A	v_A^1	v_A^2	...	v_A^m
Prod. B	v_B^1	v_B^2	...	v_B^m
...	...	...	...	...
Prod. N	v_N^1	v_N^2	...	v_N^m

As a special feature, the model considers consumers to behave boundedly rationally with only limited information about product alternatives. Based on the work of experimental economist and the bias literature (Tversky, Kahneman 1981, Kahneman et al. 1974) consumers in the model are assumed to choose products as follows:

- ▪ At the beginning consider all options that may potentially be chosen.

- Choose one characteristic of the m characteristics available.
- If one single option scores highest in respect of that characteristic, take this product.
- Otherwise, if more than one option scores similarly in respect of the adopted characteristic, remove the options with values lower than the maximum, and restart from step 2.

For example, the paper (Valente 2012) analyses two main scenarios representing different assumptions about how consumers reach their purchasing decisions. Based on the different results from the two scenarios the author concludes that the demand side and especially the consumer's behaviour should be given as much relevance as the supply side in describing market properties. However, although the model incorporates a fascinating approach to consider the multidimensional characteristics of products, it oversimplifies the role of firms and their innovation because it does not consider the knowledge level of firms.

Another interesting approach which analyses the importance of heterogeneous consumers is the model by Zakaria Babutsidze (2015). In his paper he explains a simple model of boundedly rational firms in which the heterogeneity of a firm's market knowledge, R&D behaviour and firm size arises endogenously. One key element of this model is his unique representation of demand, which is also based on the work of Lancaster (1966). The model by Babutsidze considers that consumer preferences are located in a taste space, i.e. a uni-dimensional periodic lattice – Salop's circle.

His results show, depending on parameter constellations, three regimes in which optimal behaviour of a typical firm is qualitatively distinct: (i) no R&D, (ii) R&D only on familiar markets (no diversification) and (iii) R&D on unknown markets (diversification). The market in the model also shows interesting features of the equilibrium firm size distributions. The resulting firm size distributions are fat-tailed and positively skewed thereby reproducing interesting stylised facts (see also Gibrat 1931, Lucas Jr 1978, Pavitt et al. 1987).

Finally, the so called TEVECON model by Paolo Saviotti and Andreas Pyka is an agent-based model which directly consider demand-side effects and has been applied to different scientific issues (see for example Saviotti, Pyka 2004, 2012, 2013b, 2013a). The TEVECON model is based on the work of Saviotti (1996) and represents an endogenous growth model in a neo-Schumpeterian fashion. The key characteristic of the model is that the economic system is composed of a number of sectors which are created endogenously in the model. New sectors emerge due to the innovation dynamics created within the existing ones. An interesting model extension is presented in (2012, 2013b, 2013a) in which the focus of the models lies on the co-evolution of demand and supply and the qualitative change within industries. The model considers different preference systems of consumers and shows for example that a progressive preference system results in higher rates of growth of income and of employment than conservative preference systems and that growing wages and growing levels and intensity of human capital foster long run economic development.

3   The New Agent-Based Paradigm in Economics

The concept of evolutionary economics offers a suited theoretical framework to study innovation and to understand and analyse the complex dynamics of the role of demand for innovation processes. However, it also makes high demands for the modelling framework used. This chapter describes agent-based modelling (hereafter ABM) and its particular role for the scientific endeavour.

To acknowledge the novelty of this approach, we start with a detailed introduction to the main features of ABM describing the three methodological pillars: *modelling*, *agents* and *simulation*. These pillars are common to all ABMs and frame a basic understanding of this modelling approach. In a second step, important methodological aspects vital for the successful implementation and use are discussed. In the final part of this chapter, the main implications for the simulation models are summarized.

3.1   Three Pillars of ABM

Despite the awareness that ABM has gained in the last decades, a common protocol for this method is still missing. Driven by the increasing computer resources, different scientific disciplines discovered the versatile possibilities offered by this modelling approach. This leads to different notions and understandings of this approach used in economic science.

To illustrate the diversity of notions and emphasis, ABM is also labelled as: *agent-based simulation modelling* (Polhill et al. 2001), *multi-agent simulation* (Ferber 1995, Gilbert, Troitzsch 2005), *multi-agent-based simulation* (Edmonds 2001), *agent-based social simulation* (Doran 2001, Downing et al. 2001), *individual-based configuration modelling* (Judson 1994), *multi-agent systems* (Bousquet, Le Page 2004), and *agent-based computational economics* (Tesfatsion 2002).

Regardless of a missing consent concerning the appropriate notion for this scientific method, the understanding of ABM between the different roots is very similar (see also Hare, Deadman 2004 on this issue). In the following section, we describe three important pillars of the ABM approach: *modelling*, *agents* and *simulation*, which outline the nature of this modelling approach and give a basic understanding of the model elaborated in the later chapters of this thesis.

3.1.1   Modelling from an Agent-Based Perspective

The most important element of ABM is its bottom-up perspective - describing a system from the perspective of its constituent units, i.e. the agents (Bonabeau 2002). In short, building models from the bottom-up means, letting complex macroscopic systems emerge from the interactions of microscopic entities (Epstein, Axtell 1996, Axelrod 1997). A good example illustrating the underlying principle is the artificial life program BOIDS by Reynolds (1987), which reproduces the complex behaviour of a flock of birds.

Instead of considering the flock as a self-contained unit, Reynolds (1987) was able to recreate the macroscopic behaviour by disaggregating on to the level of the system building entities, i.e. birds, and therefore building the model from the bottom-up. Interestingly, Reynolds managed to display a plausible flock behaviour, using only three simple rules for movement behaviour of the birds:

- Separation - avoid crowding neighbours (short range repulsion),
- Alignment - steer toward average heading of neighbours and
- Cohesion - steer toward average position of neighbours (long range attraction).

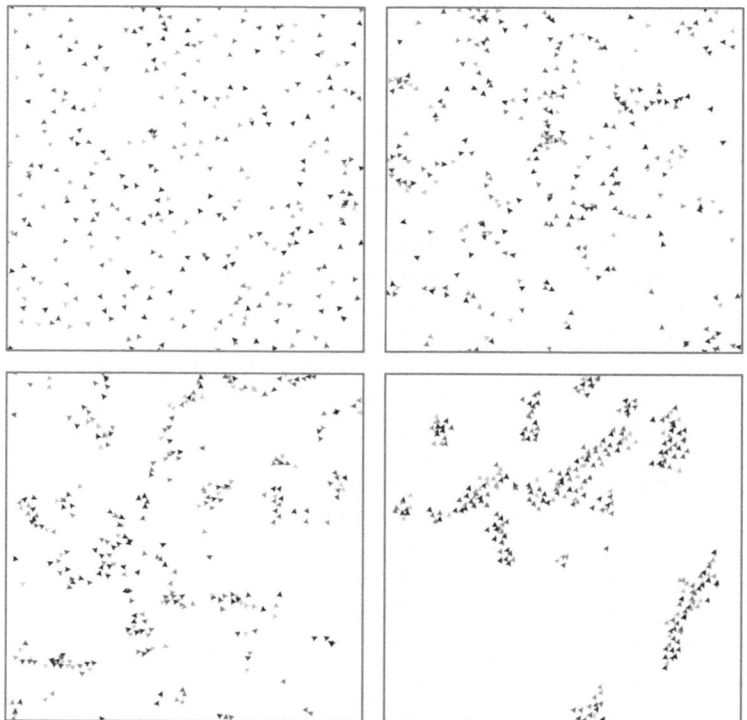

Figure 7:   Flock of Birds Created by the BOIDS Algorithm at Different Points in Time (Source: own illustration).

Figure 7 visualises the result of these simple rules, implemented in an agent-based simulation software. The triangles represent birds moving on a two dimensional grid following the basic behavioural rules found by Reynolds. Although initially distributed randomly on the grid (see upper left corner), the birds start to form coherent and supposedly self-determined flocks within only a few iterative steps (upper right corner). These first subgroups, finally, start to group together (lower left corner) so that in the end, only one flock of birds remains (lower right corner).

Although this example could be considered trivial[14], it shows the unique chances and possibilities of this modelling approach. The behaviour of complex systems as such can only be reproduced adequately by taking into account the individual behaviour, i.e. the decisions, actions and interactions of the system building units. This also illustrates how ABM differs from another numerical method which enjoys a certain popularity in economics, namely system dynamics (see for example Milling 1996, Milling 2002). On a systemic level, the rich patterns of possible dynamics remain hidden, whereas it is in the explicit focus of ABM and its bottom-up approach (Kiesling et al. 2012).

In economics, the ABM approach follows the same idea as Reynolds proposed, modelling (macroscopic) systems e.g. economic systems through the actions and interactions of (micro-) entities e.g. firms, universities, consumers etc. With this ABM departs from the top-down perspective of mainstream economic models. Instead of being constrained through representative individuals and by strong consistency requirements associated with equilibrium and an Olympic rationality, ABM describes heterogeneous entities living in complex systems that evolve through time (Windrum et al. 2007).

3.1.2   A Definition of Agents

Although the bottom-up perspective of ABM is without doubt a key characteristic, there are other approaches following the same logic of reasoning. However, in contrast to the related family of Cellular Automates (Wolfram 1986) or Microsimulations (Orcutt et al. 1986) which also exhibit a bottom-up perspective to some extent, the ABM approach places central a representation of agents in a more realistic way. Focusing on the individual behaviour of economic actors, ABMs can display important concepts such as true heterogeneity and boundedly rational behaviour of agents which leads to all sorts of variation in their modelled behaviour.

---

14   The same simulation can also be interpreted in a different, more economic, fashion: let us assume the two dimensional grid represents the possible knowledge space of firms and firm's position in this knowledge space are indicated through the triangles. Firms now move through the space via conducting R&D and follow the same rules (Separation, Alignment, and Cohesion) and create clusters and regimes of firms with similar behaviour.

An agent represents a dynamic entity, which can be assigned by the modeller with an individual role exhibiting a variety of characteristics. Although a final consensus has not been reached, in the literature it is often claimed that agents may possess the following properties (Wooldridge, Jennings 1995):

- Autonomy: Agents are autonomous entities with little or no central direction and have control of their actions.
- Social ability: Agents can interact with other agents (e.g. receiving or sending information about locations or other internal states of others).
- Reactivity: Agents have a perception of their environment (e.g. the landscape they are in).
- Pro-activeness: Agents exhibit goal-directed behaviour, taking the initiative.

The list of possible characteristics of agents can be extended in several ways. Already in their original paper, Woolridge and Jennings (1995) named more human characteristics, for example: knowledge, belief, intention, obligation and emotions etc., as possible additional features of agents (see also Jennings 1998). However, it would be dangerous to use any set of possible characteristics for us to define what ABM is and what not. For the definition of ABM, it is sufficient to say that agents are heterogeneous and autonomous individuals.

From an ABM perspective, economic actors are what they are, i.e. autonomous and heterogeneous entities embedded in an environment which is created by the actions and interactions of these agents (Gilbert, Troitzsch 2005). In contrast to traditional modelling approaches the variety of agents and their behaviour are not restricted to fit into an analytical framework (Fagiolo, Roventini 2012). Depending on the problem under investigation and the scope of the corresponding model agents can flexibly represent any kind of economic actor. On an aggregated level, this can be firms, universities, governmental bodies etc.; on an individual level, agents can be employees, scientists, consumers, households etc. From an abstract point of view, however, basically any independent component of a system can be considered as an agent (Bonabeau 2002, Macal, North 2005).

It is important to emphasize that with ABM we can specifically relax unrealistic assumptions about the agents and their behaviour. In most mainstream models, strong assumptions (for example *representative agents*) are required in order to guarantee in principle an analytical solvability (Farmer, Foley 2009). In ABM, every agent is endowed with an individual set of initial states, which allows for the representation of characteristic features and a representation of individual behaviour. In particular, agents within an agent-based model can be assumed to have only limited information about the environment and the behaviour of other agents and limited foresight about the scope of decisions or other resource limitations, such as memory, etc. (Edmonds 1999). Building on that, ABMs are capable of displaying true heterogeneity of agents (Macal, North 2005, Gilbert 2008). Heterogeneity in this sense means that agents are modelled as individual entities with individual states, and also with individual behaviours. Heterogeneity in the model can be assigned by the modeller according to the requirements of the problem under investigation, i.e. agents may be endowed with different levels of resources, initial knowledge stocks, strategies, reference systems etc. Additionally, heterogeneity is endogenously created within the model through the actions and interactions of agents themselves.

Second, as the behaviour of agent-based models is not restricted to obtain an analytical solution, agents may be assumed to behave as rationally-bound entities (Pyka, Fagiolo 2007, Gilbert 2008, see also Das 2006 for a detailed discussion). As the ABM approach focuses directly on the individual, it allows for an intentional non-rational design of economic decisions which e.g. allows for experimental adaption and learning. This enables us to model the effects of psychological principles as for example reference dependence, loss aversion and non-linear probability weighting postulated by the famous prospect theory by Kahneman and Tversky (1979).

However, the complexity of a model with individual and heterogeneous agents quickly reaches a level where for example the traditional analytical framework fails to offer any solution. A solution to this problem are computational simulation environments where even complex models can be studied in detail.

3.1.3 Simulation as In-Silicio Laboratories

As a third pillar in our understanding of the ABM approach, we have to consider that ABMs often are implemented within a computer simulation environment. Seeing simulation as a form of quasi-experiments, in principle, simple ABMs can be carried out without the help of simulation tools. A prominent example for this is the famous Segregation Model by (Schelling 1969, 1971), which was originally conducted on a chessboard using coins of different colours.

The complexity of any model grows exponentially with the magnitude of the model's assumptions quickly reaching a level where computational support is necessary. Especially through the steady improvements in computer performance, but also in the progress made on the software side (e.g. object-oriented languages and simulation environments especially dedicated to ABM such as NetLogo, LSD - Laboratory for Simulation Development (Valente 2008a), Repast - Recursive Porous Agent Simulation Toolkit etc.), today's simulations act as flexible laboratories where agent-based models can be created and systematically studied (Pyka, Fagiolo 2007).

An example of the NetLogo simulation environment (see Wilensky 1999) used for the agent-based model of this dissertation can be seen in Figure 8. The right side of the figure depicts the programming environment in which the modeller can define the model's procedures. On the left side the so called *Graphical User Interface (GUI)* of Netlogo is shown. This interface is one of the unique features of NetLogo and is designed to control basic functions of the model and simultaneously allows us to easily observe the model's output via plots.

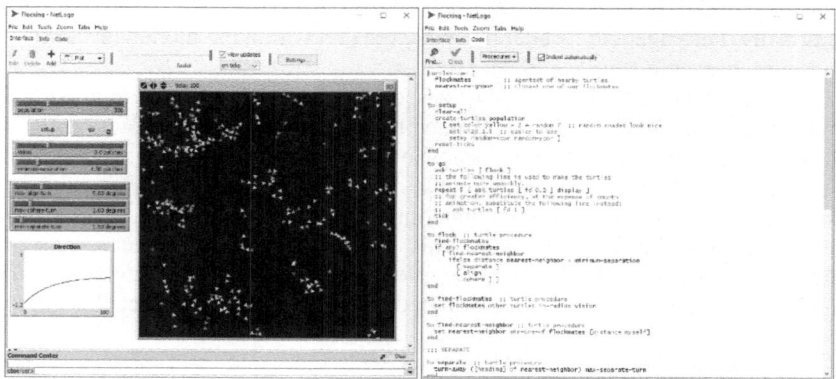

Figure 8:   NetLogo's Graphical User Interface.

NetLogo's GUI allows us to easily alter important parameters, while at the same time analysing the models output. However, simulation models often become too complex to get a comprehensive understanding of the model analysed. For this reason, for the final analysis of a model the so called behaviour space tool is applied. Here the modeller can define experiments which are then conducted systematically by the behaviour space tool. The resulting data, which can end up to contain gigabytes of information, can then be analysed through analysing tools such as Excel.

As Axelrod (1997, p. 5) puts it: "Simulation is a way of doing thought experiments. While the assumptions may be simple, the consequences may not be at all obvious." By building an agent-based model within a computer simulation environment, we have a tool at hand, which helps us to systematically observe and analyse the complex dynamics created by the actions and interactions of agents both on a macro as well as on a micro level. With a computational simulation, we are able to observe, store and analyse in detail all relevant information of the simulation as it progresses, both on a micro, but also on a macro level (Wilensky, Rand 2015). In contrast to real world experiments, simulations offer the possibility of recreating and repeating experiments with the same initial or changing conditions. This gives us the opportunity to systematically alter model parameters and assumptions to get to a comprehensive understanding of the model's outcome. A standard procedure, or to put it another way, a set of idealised steps is shown in Figure 9.

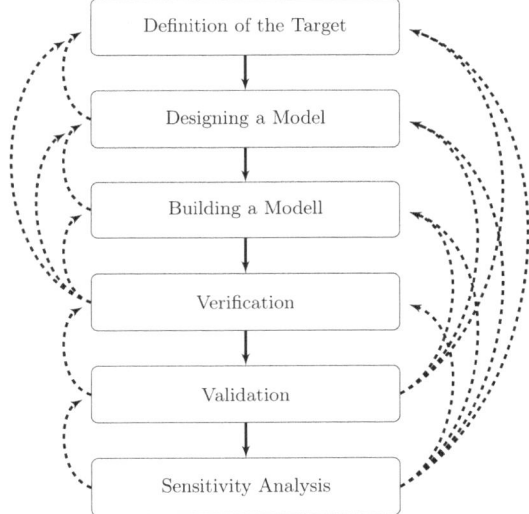

Figure 9: Idealised Steps of Using Simulations (Source: own illustration based on Gilbert, Troitzsch 2005).

Using simulation usually involves the six steps: definition of the target, designing a model, building a model, verification, validation and finally the sensitivity analysis (Gilbert, Troitzsch 2005). We start with a simple definition of the simulation's target. Based on this, we can begin to formulate a thought model, which is in a second step translated into a simulation program (*Building a Model*). After this, the modeller needs to ensure that the thought model has been implemented correctly which is called the *verification* step. In a simulation program, already small mistakes (often called *bugs*[15]) can end up to unintended results which will distort the simulation results. If the model has been finally adjusted correctly, a common step is to check whether the models output and behaviour corresponds to the intended behaviour of the thought model (see also

---

15   The term *bug* refers to an error, flaw or failure computer program which produce incorrect result. Although the origins can be traced back to Thomas Edison the first known reference of a bug to a computer problem has been made by Grace Hopper (Shapiro 1987).

subsection 3.2.4). Finally, the sensitivity analysis is the systematic investigation of different model settings and the corresponding behaviour of the simulation.

As also indicated by the dashed feedback loops in Figure 9, the term *steps* for the previously discussed elements fails to describe the iterative character of the process. Using simulation is never a clear stepwise process. In contrast, using simulations is a dynamic process in which no clear order can be maintained. So for example, sometimes the modeller realises only during the sensitivity analysis that the simulation code still includes bugs (maybe because they only apply in special cases). Another example are the various feedback loops which actually are wanted and present one of the great advantages of using simulation: it's flexibility.

Building a simulation is an iterative and evolving process in which, for a better understanding of the complex behaviour of the simulation model, often several *versions* of the model are created. The modeller starts for example with a baseline scenario in which processes within the simulation are implemented only on a very basic level. After the baseline scenario of the simulation model has been studied in detail (encompassing all elements shown in Figure 9) the modeller can then extend the scope of the simulation layer for layer (see also subsection 3.2.2).

However, the complexity involved still limits the possible scope of ABM. Although today's computer performance allows for models with an unforeseen range, the complexity of models will always be limited by the capabilities to process the data obtained by the simulation, especially on the researchers' side. In more detail, while following Moore's Law the computing power will increase constantly (Moore 1965), it remains a question whether we can use the almost infinite amount of data provided by the simulation. In fact, the bottleneck for the analysis today has become one on the user side and not on the computational side.

3.2 Using ABM as a Scientific Tool

The previous discussion on the three pillars of ABM leads us to a sound understanding of the main characteristics of agent-based models. Resting on this, we can define ABM as an approach which enables us to purposefully model complex systems from the bottom up, i.e. based on the decisions, actions and interactions of heterogeneous agents.

Despite the new and promising perspective ABM offers, it is also necessary to deal with the question of how ABM can contribute to the scientific endeavour. If used properly, ABM offers researchers both a new scientific method and a new perspective on the complex interplay between actors within an economic system. In the following section we discuss the main benefits of the ABM approach, focusing also on the main bottlenecks and pitfalls any agent-based modeller should be aware of.

3.2.1 Why Do We Need Agent-Based Modelling?

While ABM is already widely accepted in other scientific disciplines (see for example Caplat et al. 2008, Tang et al. 2014 for interesting applications within the field of biology), ABM in social science, despite gaining momentum in the last two decades, is still sometimes somewhat disputed. Accordingly, the question which needs to be clarified is what are the particular benefits of using ABM?

Especially in the field of evolutionary economics, ABM has been recognized as one of the most promising tools to investigate and to capture the dynamics and the complexity present in a model of innovation (Morone, Taylor 2010). The main objective of ABMs is to better understand the dynamic relations between micro-processes and the emergence of macro regularities. Although the particular benefits strongly depend on the intended purpose, we can identify three benefits stressed within the discussion in the literature and on which most of the agent-based modellers would certainly agree upon (see for example Bonabeau 2002, Bazghandi 2012):

- with ABM we can incorporate emerging phenomena,
- ABM provides a natural description of the real world and
- ABM is flexible.

Let us start with the first two issues. ABM as a scientific method offers a new way to analyse the complexity of economic systems. Complex systems in this sense can be defined as systems which are composed of interacting units (Simon 1995) and which show emergent phenomena (see for example Flake 1998). However, an exact definition of the term emergent phenomena seems somehow problematic (Epstein 1999). For the purpose of this dissertation we follow Epstein and Axtell defining emergent phenomena as: "stable macroscopic patterns arising from local interaction of agents" (Epstein, Axtell 1996, p. 35).[16]

Although this issue may seem insignificant, the ability to display and create emergent properties of a system based on the actions and interactions of system building entities is crucial for a sound understanding of the system's behaviour. Let us for example consider the simple case of a traffic jam. Traffic jams are the emergent result of actions and interactions of car drivers within a defined environment, i.e. streets. Already a simple model, e.g. *Basis Traffic* (Wilensky 1997) provided within the model's library of the NetLogo software shows interesting features. The model shows, for example, the counterintuitive fact that traffic jams move in the opposite direction of the cars creating it and are not only created by the number of cars but also through the myopic behaviour of drivers.

---

16   Some definitions involve the criteria of something *surprising*. However, this represents a rather vague and subjective aspect of emerging phenomena.

Based on these insights we can define several simple measures to avoid jams (Bonabeau 2002, Bazghandi 2012). Any model neglecting the agent perspective and, therefore, neglecting the main drivers of the system, clearly would have failed to give insights in such detail. In the same manner, we have to be aware that the properties of economical systems indeed are emergent phenomena (Tesfatsion 2002) created by the system building entities, i.e. firms, consumers, universities etc., and thus have to be studied from an agent perspective.

Second, ABM offers a new perspective and provides us with a more natural description of the real world. So for example, instead of defining a set of equations determining the position of each agent over time in ABM we are able to define simple behavioural rules for each agent, e.g. if [*condition*] then [*move*] or if [*condition*] then [*not move*]. This allows us to approach a more realistic representation of the respective agents and their behaviour and we can avoid the necessity to use strong simplifications to ensure the possibility to use mathematical approaches (Bonabeau 2002, Wilensky, Rand 2015). However, this issue has to be considered carefully. Although often claimed, in principle agent-based models cannot be exactly differentiated from analytical approaches using equations. Following the Church-Turing thesis every agent-based model, if implemented in a simulation environment, can in theory be implemented by a Turing machine which means that there exist a set of equivalent equations exhibiting the same behaviour (Epstein 2006). Nevertheless, it is without doubt that with ABM a natural representation of agents and their behaviour is more natural and allows for an intuitive implementation.

Finally, ABM is a flexible method. Especially if we consider the connection of powerful computers and the possibilities of simulation software which has been developed during the last two decades, ABM offers not only a new way what to analyse, it also changes how the analysis is done. With ABM we are flexible both on the input as well as the output side of the simulation. In contrast to real world experiments, simulations offer the possibility of recreating and repeating experiments with the same or different initial conditions. This gives us the opportunity to systematically alter model parameters and assumptions which allows for a comprehensive understanding of the model's outcome. Using ABM, however, also offers new possibilities to observe and analyse the models output. Within the simulation we can create a sheer infinite amount of data, observing almost every detail the simulation model provides on the agent level or on an aggregated level.

3.2.2 Managing the Complexity

Although ABM at its core makes a huge step towards a more realistic model of economic systems, one cannot expect a fully detailed picture. As with any model, an agent-based model is designed as a purposeful representation of a system rather than an exact and precise attempt to display real systems (Starfield 1990). Purposeful in the broadest possible sense can be understood that a model helps the scientist to answer questions that are of interest (Minsky 1965).

Managing the complexity within a model, i.e. finding the right level of complexity, is one of the key challenges for any agent-based modeller. As previously stated, the possibility to implement agent-based models within a simulation environment, the increasing performance of computer systems, and constantly improving methods for data analysis and visualisation enable researchers to create models of unforeseen complexity and detail. These new possibilities, however, come at a cost. As Holland (1995, p. 146) so beautifully puts it:

> "Model building is the art of selecting those aspects of a process that are relevant to the question being asked. As with any art, this selection is guided by [...] taste, elegance, and metaphor [...]."

To start, almost by definition, we will be hardly able to define something as the optimum level of complexity that should be strived for by ABM in general. Yet there is an extreme we need to be aware of: if the complexity of the model is reaching a level where we are no longer able to understand the processes involved, the experiments conducted are of no scientific use and we cannot understand these artificial complex systems any better than we understand the real ones (Gilbert, Terna 2000, Axtell, Epstein 1994). At this point the model decays to a meaningless construct without any real scientific value, which in the best case can be used to visualise our little understanding of the matter.

Despite the steady improvements in computational power, there is a second reason why the possible complexity of the model is limited. If we, for example, consider a simulation where we are for one particular scenario interested in the effects of only five different parameters, and we assume for each parameter five interesting specific values we want to analyse, we get in total $10^{10} = 3125$ different possible parameter settings[17]. If we now assume that we measure the results over 100 iterative steps and to avoid random effects let the simulation be run 100 times for each parameter setting, we arrive at 31,250,000 different data points. In other words, even with today's powerful computers and the proper software for our analysis, the analysis of an agent-based model becomes a challenging and time consuming task.

---

17   The definition of parameter of a simulation model is sometimes misleading and only based on the individual preference of the modeller and the aim of the analysis. In general, any decision and assumption in the model can be interpreted as a parameter of decisive importance, which the modeller needs to be aware of.

The level of complexity of an ABM is determined by the actors that the modeller aims to include and the number of assumptions we consider to be relevant for the model. In particular, for models of economic systems there is a broad range of possible actors and assumptions which can be relevant. The modeller must decide carefully to what extend elements of the model are necessary or negligible, facing a common trade-off: while models aiming at prediction need descriptive accuracy, models designed for explanatory power should be rather simple (Axelrod 1997).

Considering the right strategy for building the complexity within ABM, the debate has triggered a rich methodological discussion in which we find three distinct modelling strategies. First, following the KISS (*Keep It Simple, Stupid*) strategy one should start with a simple model, which may be extended if necessary. A special case of the KISS strategy is the so called TAPAS approach (*Take A Previous model and Add Something*). Here one starts with an existing model and successively complicate it with incremental additions (Frenken 2004, Pyka, Fagiolo 2007). In contrast, the KIDS (*Keep-it-descriptive, stupid*) strategy follows the idea of starting with a descriptive model first, which is then, if possible, simplified (Edmonds, Moss 2005).

An illustrative example for the KISS modelling strategy is Schelling's model of segregation in North-American cities (Schelling 1969). In this model, Schelling uses a simple grid for the representation of a city in order to model neighbourhood relationships. He succeeds in identifying the mechanisms, which lead to strong clustering patterns of ethnic groups, even if this was only mildly intended in the individual behaviour of agents (Schelling 1969).

A good example for the KIDS strategy is the model of water demand by Edmonds and Moss (2005), which includes an extremely rich set of varying behaviour rules, preference systems, water consuming devices (power showers, water-saving washing machines etc.), pricing systems, and policy options. With the help of this rich set of elements, backed by empirical observations, the authors manage to model water demand in a region close to the real water demand.

It is important to understand that, in general, the KISS and the KIDS strategies do not differ concerning the degree of complexity. In principle, it is rather a question of how to get there, although in reality it is probably inevitable that the choice for a strategy concerning building the relevant assumptions will end in models with considerably different levels of complexity. Without going into too much detail, the debate has many facets and a set of additional strategies must be included for a full picture (Pyka, Fagiolo 2007). It is, however, important to note that agent-based research should not be oppressed by the fear of complexity. Any claim that ABM is limited to investigate only simple dynamics in small systems is neglecting the possibilities of ABM. Applying ABM for the study of economic systems requires a detailed methodological understanding of this research method and additional inputs from other disciplines, i.e. through a profound understanding of realistic behavioural heuristics (see for example Gigerenzer, Selten 2002).

3.2.3   Two Ways of Using Agent-Based Models

As stated before, the bottom-up perspective of ABM builds on the actions and interactions between heterogeneous agents to analyse the multilevel effects on the overall system, the environment and the agents themselves. Through this, ABM offers a unique understanding of the processes of the interplay between micro and macro levels of a complex system. However, the particular role of ABM for the scientific endeavour still is under debate. Facing critiques questioning the scientific value of ABM the modeller needs to be aware of how ABM can be used to deepen our understanding. Based on the literature we can distinguish between two, as will be argued later complementary, perspectives facing different questions.

The key for understanding the basic differences between both perspectives can be seen in Figure 10. As stated before, one of the key characteristics of ABM is its bottom-up approach to reproduce the behaviour of systems. Let us consider correspondingly the micro and the macro level of economic systems. In section 3.2.1 we explained that through its bottom-up perspective ABM can capture emergent phenomena, i.e. "stable macroscopic patterns arising from local interaction of agents" (Epstein, Axtell 1996, p. 35). These patterns on the macro level create also the environment for the actions and interactions of the system building entities on the micro level and, in turn, changes the conditions under which the macro level patterns have been created. In other words, in ABM we can capture both levels and the mutual relationship between both levels in which the clear cut between *cause* and *effect* diminish.

Nevertheless, in the literature on ABM two different ways of thinking emerged. One the one hand, we can start with known macro level phenomena looking for micro specifications creating these patterns. On the other hand, we can also start with interesting micro level specifications and analyse the corresponding output on the macro level (Figure 10).

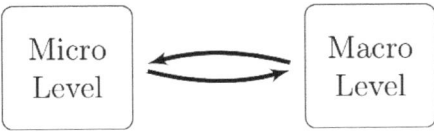

Figure 10: The Micro and Macro Level of ABM.

Following the idea introduced by Epstein and Axtell (1996), ABM can be used in a generative perspective.[18] ABM in this sense focuses on the possibilities to display the emergence of complex macroscopic system behaviour by the actions and interactions of agents on a micro level. In other words, agent-based models can be designed to find micro-specifications that can explain (from a generative perspective: *grow*) a macro-level phenomenon of interest (Epstein 1999). In the literature this is often put on a level of reproducing stylised facts through ABM.

To name just one example for a fascinating model reproducing stylised facts, in 1996 Epstein and Axtell (1996) reproduced right-skewed wealth distributions, amongst other things, based on their seminal Sugarscape model. The scope of the model has been continuously extended during the past years, implementing new features carrying forward new aspects to the model (see Epstein 1999 for a list of other interesting models reproducing stylised facts).

The logic of reasoning behind the generative perspective of modelling stylised facts constitutes also a severe methodological caveat of the ABM approach which is stressed also by Epstein (1999) or Gilbert and Terna (2000): In principle it is possible that similar macroscopic system behaviour might be generated by models which refer to different assumptions with respect to the behavioural rules of agents on the micro-level. The agent-based modeller simply cannot presume that the microscopic behaviour of agents found to reproduce stylized facts is in fact an accurate and relevant description of the phenomena of interest. For this reason, a final proof for the validity of an agent-based model which reproduces stylized facts is only hard to be given.

Using agent-based models, however, is not limited to reproduce stylized facts in order to find possible explanations for emerging macroscopic patterns. While the starting point for ABMs following the generative idea are patterns on the macro level, another strand of models in the literature is based on the possibility to perform a wide range of numerical experiments. These models aim to analyse what emerging patterns arise based on particular micro specifications of the model. The focus here is on the improvement of our understanding of the dynamic processes within a complex system. ABM from this perspective acts as a laboratory for computational experiments created *in-silicio* (Pyka, Fagiolo 2007, Leombruni 2002). As also Axelrod (1997, p. 5) puts it:

> "Simulation is a third way of doing science. Like deduction, it starts with a set of explicit assumptions. But unlike deduction, it does not prove theorems. Instead, a simulation generates data that can be analysed inductively. Unlike typical induction,

---

18   Using agent-based models in a generative sense can also be considered as using it following the logic of abductive inference (see Leombruni 2002, Richiardi et al. 2006, see also Peirce 1901, Popper 2014 for a more general discussion).

however, the simulated data comes from a rigorously specified set of rules rather than direct measurement of the real world."

To put this another way, detached from the possibility to reproduce stylized facts, ABM can also be used as a scientific method to increase our understanding of complex systems. While the starting point of ABM in a generative sense are macroscopic patterns, the starting point of ABM as an *in-silicio* laboratory is a set of specifications for the micro level, i.e. a well-grounded set of behavioural rules on the agent level. In this laboratory, the modeller is able to observe and analyse the different outcomes of different assumptions underlying the simulation model.

Building on Tesfatsion (2006) we can identify two different objectives of ABM as *in-silicio* laboratories. First, ABM can be used to get a normative understanding, for example, evaluating whether designs proposed for economic policies, institutions and processes will result in socially desirable system performance over time. Second, ABM allows for qualitative insights and theory generation. In this case, the objective is to understand economic systems through a systematic examination of their potential dynamical behaviours under alternatively specified initial conditions.

However, using ABM in a generative way or using it as an *in-silicio* laboratory is not mutually exclusive. As we will see, using ABM allows for both aspects simultaneously. Although the main goal of an ABM might be to analyse the outcome of given micro-specifications, in some cases we find these outcomes to by identical to known empirical facts which will, in turn, to some extend also help us to validate additional findings.

3.2.4 The Need for Verification, Validation and Calibration

The most common critique against agent-based models is the perceived lack of robustness. To start, any modeller needs to be aware of the possibility of simple failures in the implementation of the model in a computer simulation. This verification of the computer program is a crucial step for the credibility of a model (see subsection 3.1.3).

Verification simply asks whether the model does what we think it is supposed to do (Ormerod, Rosewell 2009). However, verification is more than just looking for bugs in the computer code. As for example Dawid and Kopel (1998) note: "we have to be aware of the fact that simulation results may crucially depend on implementation details which have hardly any economic meaning".

In other words, even though the technical code might be correct, the way how the model is translated and implemented in a computer software might lead to biased results which have strong effects on the model's outcome, yet were unintended by the modeller. Especially for complex models, it is necessary to make the implementation of an ABM transparent for other researcher. This can either be done by providing the original simulation code or by making main procedures public, including the pseudo code in the publication.

Despite the correct implementation of the simulation model, the modeller is also often confronted with critiques questioning whether the model is an accurate representation of the real world from the perspective of the model's intended applications, i.e. the model's validity (Ormerod, Rosewell 2009). Following the distinction made in subsection 3.2.3 we can differ between two ways of validation: i.e. the input validation and output validation. While the latter refers to the matching of model results and against acquired real-world data, the former regards ensuring that the fundamental structural, behavioural and institutional conditions incorporated in the model reproduce the main aspects of the actual system (Bianchi et al. 2008).

Without going into much detail, this issue is still under great debate and several strategies have been developed to approach the problem of validity of agent-based models, such as the indirect calibration approach and the Werker-Brenner approach (Windrum et al. 2007, Werker, Brenner 2004) etc. The underlying concept of these validation strategies is an elaborate multilevel approach where input and output validation is combined. Starting with a model designed to reproduce stylized facts, the modeller can use the micro-specifications found to be valid to replicate the macro patterns as the starting point for a wide set of simulations experiments aiming to give further insights into the dynamical behaviours of the model (see for example Windrum et al. 2007, Ormerod, Rosewell 2009, Werker, Brenner 2004 for interesting discussions on this issue).

Although without doubt the validation is of particular interest, especially if we want to derive valid outcomes such as policy recommendations, we have to be aware that the complexity of ABM always invites criticism. Despite any thorough validation of the model, we may be confronted with questions whether all elements of the model are necessary or if other, so far missing elements, are relevant too and, thus, should be included. Second, the possibility of performing validation is restricted by the set of relevant data available. For validation of agent-based models we need more than some widely accepted macroeconomic stylised facts. As ABM builds on the actions and interactions of heterogeneous economic actors, this approach requires a fundamentally new understanding of the behaviour of these actors.

3.3 Implications for the Following Analysis

Summing up, shifting the focus from an oversimplifying perception of economic systems to a more realistic one, ABM is designed to overcome the limiting possibilities of a traditional analytical framework. With ABM we can overcome the strict homogeneity assumption of traditional aggregate models (Kiesling et al. 2012). Central to this new approach is its exceptional perspective of economic actors, treating economic agents as heterogeneous and individual actors which build economic systems by their decisions, actions and interactions from the bottom up.

Using this method comes at cost. Gell-Mann (1969) Nobel laureate in physics and co-founder of the Santa Fe Institute is attributed with saying: "Imagine how hard physics would be if electrons could think". Applying ABM to economic processes faces exactly this problem. We need to be aware of how ABM can be used to deepen our understanding of complex economic processes. Although ABM at its core makes a huge step towards a more realistic model of economic systems, one cannot expect a fully detailed picture. As with any model, an ABM is designed as a purposeful representation of a system. Purposeful in that sense means it can be used in a generative way, reproducing macro level patterns and, thus, finding possible explanations on the micro level, but also as an *in-silicio* test laboratory where we study the outcome of different micro-level specifications.

Especially for the latter case, despite any effort to ensure validity of our models, the complexity created within ABM invites for critiques. To encounter them, we need a better understanding of the behaviour of economic actors but also well accepted standard-models. So far, ABM cannot aim for including all possible aspects of an economy. In contrast, we need at this point to focus on a new joint understanding of basic economic processes, emerging from a profound acknowledgement of the decisions, actions and interactions of all economic actors. Building on that, we can stepwise increase the complexity of our models, gaining descriptive accuracy and, with that, increasing the predictive power of ABM.

From this follows that the agent-based simulation model carried out in the following chapters cannot aim at presenting a full picture of the demand side or innovative firms, nor do we aim to calibrate or validate our model based on empirical findings. Instead, we focus on fundamental processes to gain a first understanding of the relevant processes. Given the discussion on the state of the art in the literature on innovation and demand, this means that we first heterogeneous consumers into the analysis and investigate the implications of doing so. Building on that we can stepwise increase the complexity, adding further aspects into the model.

4 An ABM of Heterogeneous Consumers and Demand

After the discussion of the ABM approach in the previous chapter, the following chapter describes the agent-based simulation model used as a baseline model. This baseline model shows the implications of introducing heterogeneous consumers with heterogeneous demand into the analysis of innovation. Hereby, we aim at creating a computational framework which can display both, the heterogeneity of the demand as well as the supply side and their effect on the innovation process.

This chapter is organized as follows: We start in section 4.1 with some introducing remarks on the scope and the main aim of this agent-based model. In section 4.2 we then describe the basic concept of products, knowledge and consumer heterogeneity of demand in the model and provide a detailed model description. In section 4.3 we explain and analyse the results stressing the importance of including consumer heterogeneity for the analysis of innovation processes. Finally, we provide in section 4.4 preliminary conclusions.

4.1 Introducing Remarks

The following agent-based simulation model represents a baseline model, which is analysed in detail in this chapter and stepwise extended throughout the later parts of this dissertation. The main goal of this model is to introduce an agent-based simulation incorporating consumer heterogeneity into the analysis of innovation processes.

With the following agent-based simulation model we aim to incorporate consumer heterogeneity and link it with the innovation activities by firms. Inspired by the work of Kelvin Lancaster (1966, 1975, 1979), the simulation model incorporates heterogeneity of demand based on heterogeneous consumers equipped with individual preferences for the particularities of product characteristics (e.g. the particular colour or size of a product, the functionality, included services etc.). This concepts becomes more clear if we consider the following words by Lancaster (1966, p. 134):

> "Consider the choice between a gray Chevrolet and a red Chevrolet. On ordinary theory these are either the same commodity […] or different commodities [...]. Here we regard them as goods associated with satisfaction vectors which differ in only one component, and we can proceed to look at the situation in much the same way as the consumer-or even the economist, in private life-would look at it."

Lancaster stresses here the important fact that products and with that innovations are more than just vague and abstract concepts for which price and quality are the only essential aspects. In contrast, we have to consider that innovations have multiple characteristics (or *components* as Lancaster puts it). So instead of simplifying the demand for products to only two characteristics (e.g. price and quality) as many other models do, in the model the individual demand includes a full image of a preferable product within a multidimensional product characteristic space. With this we aim to introduce a general framework model of interaction between heterogeneous producers and heterogeneous consumers in an environment where producers are forced by the innovative activities of competitors to continuously adapt to market conditions.

Following the ideas of KISS and TAPAS (see chapter 3), the simulation model is kept, one may rashly say, simple. This means that we only consider firms and consumers as relevant agents for the analysis and do not consider other influences such as private research institutes and universities as additional agents or public procurement as additional factor of demand. This neither represents a comprehensive picture of the demand side nor does it include other major drivers of the innovative processes. Instead, we focus directly on the consequences of different degrees of heterogeneity of demand on the innovative behaviour of firms.

Figure 11 shows the relevant scope of the simulation model. On a firm level, we focus on the functional relationship between the individual knowledge stock of a firm and its products. Although this link is without doubt of vital importance for the analysis of innovation, only few agent-based models of innovation consider this link. An exception from this is the so called SKIN model (*Simulating Knowledge Dynamics in Innovation Networks*), an agent-based model developed by Nigel Gilbert, Petra Ahrweiler and Andreas Pyka (see for example Gilbert et al. 2001, Ahrweiler et al. 2004, Gilbert et al. 2007) which considers a flexible way of a computational link between knowledge of firms and the characteristics of the products generated (see also chapter 2.3).

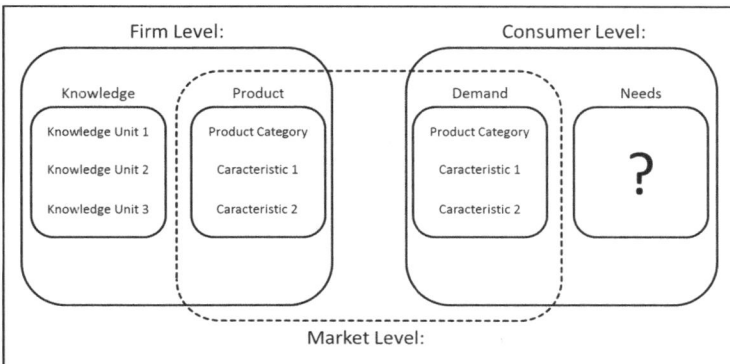

Figure 11: Scope of the Baseline ABM.

We claim that the relational link between knowledge of firms and the characteristics of products produced needs to satisfy the following properties. First, knowledge in our simulation is at least partly *substitutive,* meaning that the use of different knowledge stocks may eventually lead to the same product characteristics[19]. Second, there are possible *interdependencies* between product characteristics, e.g. it is hard to produce a product which is robust and lightweight at the same time. Third, the functional relationship between knowledge and products needs to be *transparent* as well as *variable* and *computational.* While the latter three characteristics of the relational link between knowledge and products is obvious from the perspective of an agent-based modeller aiming to analyse and understand the behaviour of the simulation model in detail, the first two characteristics result from the need to represent the innovation efforts of firms adequately.

The consumer level can generally be separated into individual needs and the resulting demand for products. For this stage of the simulation model we renounce the abstract link between something as shallow and intangible as consumers' needs and the resulting demand (see for example Godin, Lane 2013 for an interesting discussion on the difference and its consequences) and focus only on the actual demand of consumers.[20] Instead of considering only the price or the quality of

---

19   This fact can also be described by the term *equifinality*. The term equifinality was coined by
      Driesch *(1908)* and Bertalanffy (1969) and describes the aspect of systems that a given end state
      can be reached by different means.

20   Following Mowery and Rosenberg (1979) and Godin and Lane (2013) human needs are a
      shapeless and elusive notion. Needs are per definition unlimited, and therefore not capable of
      driving decisions about innovation, while demand is identifiable using precise (economic)
      criteria.

products we consider a multidimensional product characteristic space where products are considered to exhibit a large set of characteristics (Lancaster 1966, Lancaster 1975).

In a nutshell, the simulation model can be described as the numerical description of the innovative adaptation process of profit-driven firms trying to sell their products to heterogeneous consumers characterised by individual demand. Concerning the previously mentioned supposed *simplicity* of the model it is important to note that the simulation with its basic setting already took months of computing time and provided several gigabyte of data analysis. Although the following analysis of the models' output focuses on some selected parameters only, the analysis undertaken covers an almost full picture of the importance and effects of all crucial assumptions, including not only parameters per se but also basic decisions about vital simulation procedures. Even if the scope of the model does not cover a full picture of the innovation process and the demand side, the baseline simulation model still provides important results, adding to our understanding of the relevant dynamics within innovation processes.

As a simulation environment for the agent-based model we decided to use the NetLogo simulation environment made available by the *Northwestern's Center for Connected Learning and Computer-Based Modeling* (CCL) and authored by Uri Wilensky (1999). In Figure 12 we show the graphical user interface of the simulation model. The user interface includes several plots to monitor important values, a *world view* in which the consumers and firms are displayed and a set of sliders and buttons to control the simulation and to alter major parameters.[21]

---

21   Although NetLogo's graphical user interface provides all possibilities to control and the simulation and to visualize the results we also used the possibilities of the so called behavior space which lets us define experiment settings with altering parameter values and to export all relevant simulation data for the external analysis.

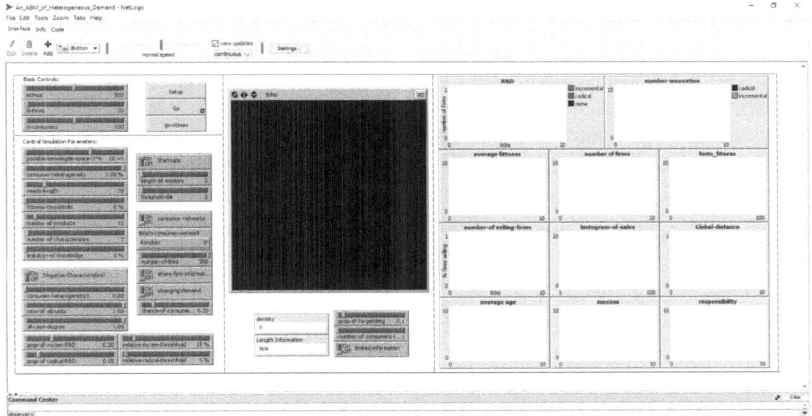

Figure 12: Graphical User Interface of the Simulation Model.

Our starting point for the analysis is a standard scenario which aims to show the fundamental impacts of consumer heterogeneity on the innovative behaviour of firms. For this purpose, we focus on the effects of heterogeneity of consumers. With this simulation we can vary consumer preferences on a spectrum between full homogeneity and full heterogeneity and, hence, analyse the innovation behaviour of firms and the resulting market structure. In a second step we use different experimental settings for *in-silicio* policy experiments, investigating the impact of different policy strategies designed for fostering innovation.

4.2   The Baseline Simulation Model

This section describes our approach to implement the heterogeneity of consumers into an agent-based simulation model. We start with a general introduction on how demand, knowledge of firms and products are modelled. In subsection 4.2.2 we describe the processes during a simulation run in more detail.

4.2.1   Modelling Multi-Dimensional Product Characteristics

Computational agents are empty vessels the modeller needs to fill with characteristics and behaviour, taking into account the strict boundaries of a software simulation environment. Building an agent-based model in a simulation environment means a careful and well-grounded translation of our thought model into programming language. Although simulation environments and the underlying programming language are continuously advancing, the modeller still faces the basic challenge to translate sometimes very abstract thought models into a viable program code. Consequently, the key elements of the following simulation model of heterogeneous demand are well-suited representations of knowledge, products and demand and their links (see section 4.1).

Following the ideas of Lancaster (1966, 1975, 1979), products in the simulation exhibit a set of characteristics forming a multidimensional characteristics space. Other models using this ideas are for example Saviotti and Metcalfe (1984), Saviotti (1996), Gallouj and Weinstein (1997) and Valente (2012). This set of characteristics A in our simulation is represented by a bit string of finite length, defining all possible combinations of product characteristics for a specific product k of a firm j. For example, in equation (4), we depict a possible product and its set of characteristics as follows:

$$A_{k,j} = 10001101010011010010 \tag{4}$$

This particular representation of product characteristics introduces a multi-dimensional product characteristic space and allows us to consider true heterogeneity of demand.

This also has strong implications for the requirements of knowledge representation. Following Gilbert et al. (2001, 2007) we begin with an abstract characterisation of knowledge by assuming knowledge $\bar{K}_j$ of a firm j to be a set of single knowledge units ($K_{i,j}$). Equation (5) shows the knowledge endowment $\bar{K}_j$ of a representative firm j with three knowledge units $K_{i,j}$:

$$\bar{K}_j = \begin{Bmatrix} K_{1,j} \; e.\, g.\, 1010110000 \\ K_{2,j} \; e.\, g.\, 0010000000 \\ K_{3,j} \; e.\, g.\, 1101100001 \end{Bmatrix} \tag{5}$$

Describing not only the product characteristic but also the knowledge units as binary bit strings allows for an elegant definition of unique mapping functions $M_{A_{k,j}}$, defining the functional relationship for any product characteristic $A_{k,j}$ for the respective product k. In other words, the mapping functions contain the information on how to read the knowledge from equation (5) in order to compute a firm's product. To clarify this, let us assume that a potential firm has a knowledge stock as depicted in equation (5). This knowledge endowment of a firm is then read according to the mapping function $M_{A_{k,j}}$ in equation (6) in order to obtain the respective product characteristic for product k:

$$M_{A_{k,j}} = 1,1,1,2,2,2,2,3,3,3 \tag{6}$$

The mapping function $M_{A_{k,j}}$ defines that the first, second and third bit of the characteristic bit string is defined by the first three bits of the first knowledge unit of a firm. The following 4 bits are then defined by the bits four to seven of the second knowledge unit and the last three bits of are defined by the last three bits of the third knowledge unit. Hence, for the product produced by the firm from equation (5), the following bits (underlined) are relevant:

$$\overline{K}_j = \begin{cases} K_{1,j}e.\,g.\,\underline{101}0110000 \\ K_{2,j}e.\,g.\,001\underline{0000}000 \\ K_{3,j}e.\,g.\,1101100\underline{001} \end{cases} \qquad (7)$$

With a mapping function as given in equation (6), this leads to the product characteristics $A_{k,j}$ as follows:

$$A_{k,j} = 101\ 0000\ 001 \qquad (8)$$

The definition of the functional relationship between knowledge and products via mapping functions allows us to easily compute a large set of different products and product characteristics in a transparent and viable way. Additionally, we see that through the use of mapping functions, the use of knowledge is to some extend substitutive as only some parts of a knowledge unit are used to define a particular product characteristic. However, as some information of a knowledge unit can be defined as relevant for more than one product characteristic, the use of mapping functions also mirrors the interdependence of different product characteristics.

Bit strings are used not only to describe knowledge units or product characteristics but also to equip consumers with an individual demand. More precisely, each consumer n is assigned an individual demand $D_{k,n}$ for all products n including a full picture about the desired characteristics:

$$D_{k,n} = e.\,g.\,0011010010 \qquad (9)$$

With this we can easily model the full spectrum between homogeneous demand (the desired product characteristics are the same for all consumers) and full heterogeneous demand (every consumer has an individual demand for products). To be able to also model the spectrum in between these two extremes, we define $p \in [0, \dots, 1]$ as a global parameter to determine the level of heterogeneity of demand in the simulation. Starting with a randomly defined global standard demand for each product and product characteristic, p is the probability for each bit of the demand sequence to be reset for each consumer randomly. With this we can easily switch between the two extremes: homogeneous markets ($p = 0$; i.e. all consumers follow the standard) on the one hand and fully heterogeneous markets ($p = 1$) on the other hand. In this situation every consumer has its individual product in mind which he or she wants to buy. Additionally, also in-between degrees of heterogeneous markets (e.g. $p = 0.5$ where consumers share only parts of the bit sequence) can be analysed.

To measure the effective heterogeneity in the markets we define the relative distance between consumers as the pairwise comparison of all consumer preferences. Figure 13 shows the resulting relative distance for different values for our heterogeneity parameter p. As a consequence of the large number of consumers and the length of the product characteristic space, the maximum difference between consumers on average never exceed a 50% match. Additionally, the figure shows that for the effect of an increasing heterogeneity parameter p for high values of heterogeneity diminishes.

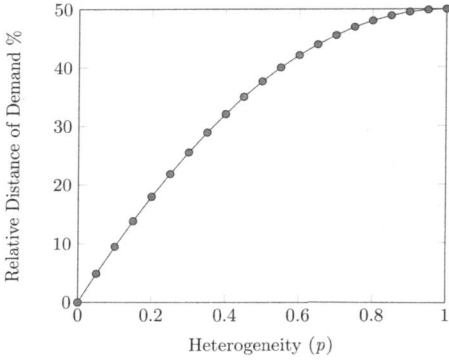

Figure 13: Relative Distance of Demand for Different Levels of p.

4.2.2   Basic Procedure of the Simulation Model

A simulation run can be generally divided into the initialisation phase and the repetitive run of the simulation procedures. Figure 14 shows an overview of these procedures: each time step, firms try to sell their products to consumers with each product being characterised by a number of product characteristics and a number indicating the product category. Based on market returns (i.e. the number of sales made by a firm), firms decide to engage in research and development, i.e. to innovate by changing their knowledge stock and hereby adapting the product to the market demands. Finally, a constant number of firms enter the market and unsuccessful firms leave the market.

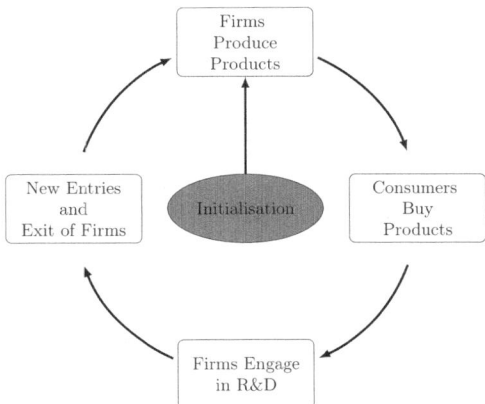

Figure 14: Iterative Steps of the Simulation.

During the initialisation phase, we first populate the simulation with a defined number of firms and consumers. Each firm then is endowed with a random set of three knowledge units (see equation (5) for an example) and each consumer is equipped with the preferences for a fixed, individually optimized product. Finally, the necessary mapping functions are generated (see subsection 4.2.1). Additionally, the simulation creates one (global) mapping function to determine the specific product category a firm's product is assigned to. For this, we simply compute the decimal value of the resulting bit string and divide it modulo the absolute number of products. The procedures for the rest of the simulation are implemented as follows:

- Each firm produces a product based on its individual knowledge stock
- Consumers analyse all products available, purchasing the product best matching the consumers' preferences
- Firms engage in R&D in accordance with their innovation strategy
- Firms without sales for more than five periods exit the market and a constant number of new, randomly generated firms enter the market

To give an example of how a firm's product is defined and how consumers evaluate products, let us assume a random firm j to be endowed with the following knowledge stock (equation 10):

$$\bar{K}_j = \left\{ \begin{array}{l} 0110011011 \\ 0010001111 \\ 1010101100 \end{array} \right\} \tag{10}$$

Furthermore, let the mapping function to define the product category M_g be:

$$M_g = 1,1,1,2,2,2,2,3,3,3 \tag{11}$$

This leads to the temporary bit string:

$$011\ 0001\ 100 \tag{12}$$

Written in decimals this bit string equals 792, which then defines the product category 2 considering 10 possible product categories ($792\ modulo\ 10 = 2$). Finally, let us assume that products are characterised by one set of product characteristics and the respective mapping function M_1 as:

$$M_1 = 2,2,2,1,1,1,1,3,3,3 \tag{13}$$

Which defines the product characteristics $A_{1,j}$ as:

$$A_{1,j} = 0010011100 \tag{14}$$

After the production of products, consumers choose and buy the – according to their perspective – optimal product. Obviously, a full fit between a product offered by a firm and the product desired by a consumer can only rarely be observed. In order to determine the match between characteristics consumers demand and the characteristics firms offer (i.e. the consumer preference match, henceforth CPM), the inverted Hamming distance (Hamming 1950) is applied; i.e. first it is determined on how many positions over all characteristics of the product the product offered differs from the product desired (see equation 15). This number is then divided by the maximum number of digits.

$$\Delta(A_{k,j}, D_{k,n}) := \left| A_{k,j} \neq D_{k,n} \right| \tag{15}$$

Equation (16) shows a simple example on how the match between a consumer's demand and a firm's offer is determined. In this example, the six underlined bits match the demand of the consumer. In total there are ten bits which leads to a CPM level of 60%.[22]

---

22   In case two products feature the same CPM level consumers first check whether one of these firms is also the last period's seller. If so they keep buying their product from this seller, if otherwise they choose randomly between the products.

$$A_{k,j}: \{00\underline{10}0\underline{11}\underline{100}\} \Leftrightarrow \{1110110100\}: D_{k,n} \qquad (16)$$

In a next step, firms decide about their R&D strategy. The innovation process in our simulation is a highly uncertain process which involves the (re-)combination of knowledge (Dosi 1988b). A firm's engagement in R&D is based on two criteria, both of which are conceptualized as thresholds within the firms' decision routines (see for example the models in Nelson, Winter 1973, 1982). First, we define a firm's market share φ_j as firm j's share of consumers having demand for products in the respective product category of firm j. If the market share is low (below the threshold for radical research h^r) the firm will engage in radical R&D. This means that a firm is deleting one of its knowledge units entirely replacing it with another knowledge unit randomly chosen. Radical R&D therefore can be interpreted as the attempt of a firm to search for new market niches since the potentially new characteristics will be perceived very differently on the consumer side.

Furthermore, for cases in which the firm's market share is above the radical threshold but still below the threshold for incremental research h^i, the firm will engage in incremental R&D. The respective firm only changes one bit of one of its randomly chosen knowledge units, trying to more successfully adapt to the needs of consumers in an existing market niche. With a firm's market share being high enough (i.e. above the incremental threshold) the firm no longer engages in R&D. This means that the respective firm is satisfied with its market position avoiding the risk of possible negative consequences of modified knowledge about the characteristics of its products.

Finally, our model reflects Schumpeterian competition based on innovation, leading to creative destruction by allowing for entries and exits (see for example Nelson, Winter 1977). We assume that each time step a constant number of new firms enters the market in an entrepreneurial fashion. Simultaneously, firms are forced to exit the market if they are unable to sell their products for more than five consecutive periods.

4.3 Simulation Experiments

The following description of our simulation results is subdivided into two set of experiments: First, in a standard scenario (subsections 4.3.1 and 4.3.2) we stress the dynamic effects on innovation caused by the heterogeneity of consumers. With this we aim to show the fundamental dynamics caused by covering the fact that products are more than goods which can be characterised only by price and quality. Finally in subsection 4.3.3 experiments show a possible application of our model investigating the impact of different policy strategies intended to foster the performance of an economy via innovation.

4.3.1 Innovations in a Multidimensional Characteristic Space

In our first scenario we use the parameter setting and initial values displayed in Table 2. Although we focus in our first analysis on the effects of the heterogeneity parameter p, other parameters such as the number of consumers, the complexity of product characteristics or the particular R&D thresholds are also of vital importance and have been analysed intensely. However, the parameter setting shown in Table 2 proved to be robust for the first set of experiments to analyse the basic processes within a heterogeneous demand environment. Additionally, unless otherwise stated, for each parameter setting the simulation is run 1000 times to minimize the random effects and average results are shown.

Table 2:   Parameters and Initial Values of the First Experiment.

Number of Consumers	100	Initial Number of Firms	20
Number of Product Categories	10	Length of Knowledge Units	10
Product Characteristics	1	Length of Product Characteristics	20
Market Share for Incremental R&D	< 30%	Market share for radical R&D	< 10%
Chance of Incremental Innovation	30%	Chance of Radical Innovation	15%

We start with a simple analysis of the adaptation process of firms throughout a single simulation run. Based on the market feedback, i.e. the number of sales, firms decide to engage in R&D to change their knowledge stock. Over time, this leads to a general adaption and improvement of products to demand. Figure 15 depicts this adaptation process over time for a market of homogeneous and heterogeneous demand. As the figure shows, in both cases firms adapt their products successfully to the demand which leads to an overall increase in the CPM level. Second, the results show that over time these improvements lose on magnitude, reflecting the increasing difficulties to achieve successful innovation.

However, the most important insight can be obtained if we compare the development for homogeneous and heterogeneous demand in more detail: while the improvements made in a homogeneous market seem to be clearly stepwise, the progress made by firms facing heterogeneous demand is more smooth and also exhibits sometimes a temporally decrease in the CPM level. The explanation for this difference is the dynamic segmentation of the market in case of heterogeneous demand.

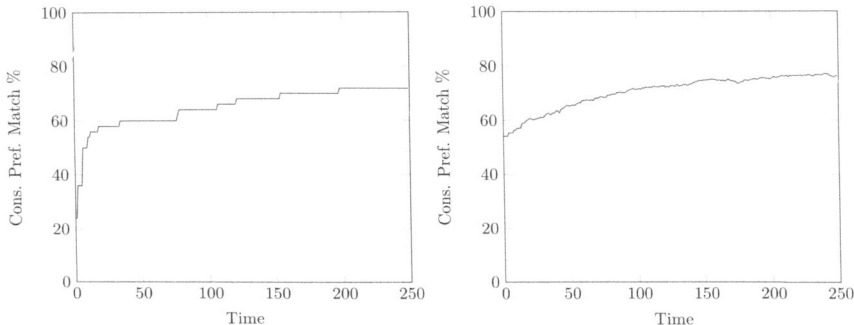

Figure 15: Average CPM Level for Homogeneous (left) and Heterogeneous (right) Consumers.

Starting with a case of homogeneous demand ($p = 0$), all consumers share the same demand for product attributes. Because consumers are fully informed about all possible products and we do not consider limited information of the consumer side, the market clearly divides into market leaders serving all consumers in a specific product category and follower firms who do not have any consumers. In this situation the market leaders can be considered as temporal monopolists. The followers try to overtake the market by engaging in radical R&D and, as a consequence, technological progress only occurs stepwise. If an innovation of a firm creates a better product (matching the demand better than the product of the existing incumbent) the innovative firm immediately becomes the new market leader and the CPM level increases. However, as consumer preferences are fixed, the possibilities for new and better solutions decrease over time and eventually are exhausted when a firm discovers a knowledge combination that leads to a product perfectly matching the consumers' preferences.

For the case of heterogeneous demand, the market divides into several segments of consumers. An innovation of a firm in this case may improve the product from the perspective of some consumers. Others, however, may consider this innovation as negative which makes the new product in some cases irrelevant. As a result, the progress made by firms is characterised by only small steps smoothening the overall development and sometimes even leading to an on average decrease of CPM levels. The key difference between the two extreme cases shown above, can be explained best by the difference in the resulting fitness landscape generated by the multidimensional characteristic approach introduced in this agent-based model. We use the reference of a fitness landscape following the famous NK-Model by Stuart Kauffman to refer to the search problem firms face confronted with the multidimensional product characteristics and heterogeneous demand (Kauffman, Levin 1987, Kauffman, Weinberger 1989, Levinthal 1997, Fleming, Sorenson 2001, Fleming 2001, Frenken 2006).

The NK-Model was originally developed by Kauffman in a biological framework. He coined the term *Tuneable ruggedness* which captures the intuition that both the size of the landscape and the number of its local maxima can be adjusted by changes to its two parameters, N and K. The model has found use in many fields including in the study of epistasis and pleiotropy in evolutionary biology, organisational theory, management, complexity science and combinatorial optimisation. However, we find also several attempts to modify and extend the original model to adapt the model to a more economical approach (see for example Valente 2008b).

To get a better understanding of this problem let us reduce the search problematic of firms to only two possible product characteristics and assume firms' products can be visualized by two dimensions (Characteristic A and B in Figure 16).

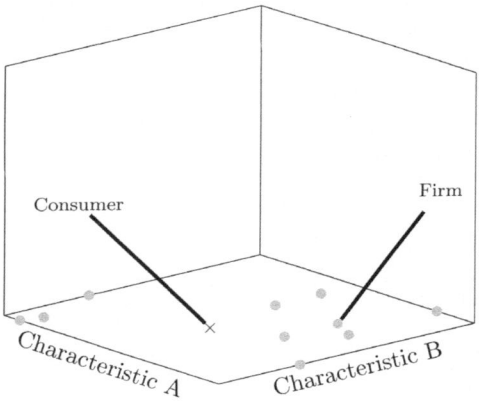

Figure 16: 2-Dimensional Search Problem with Homogeneous Demand.

In case of homogeneous demand, all consumers share the preference for a particular combination of product characteristics (A,B) represented by the cross in the above figure. Firms' products are represented by dots scattered around the (A,B) surface. In order to increase profits, firms move through the (A,B) surface altering the product and eventually improving the match between consumers' preferences and their product indicated by the distance between a firm's product and the demand in Figure 16.

We can interpret the search problem for higher sales in Figure 16 - in reference of the NK-Kauffman model - as a fitness landscape as indicated in Figure

17. In case of homogeneous demand, this then results in a single peak or hill in our fitness landscape firms try to climb through innovation.

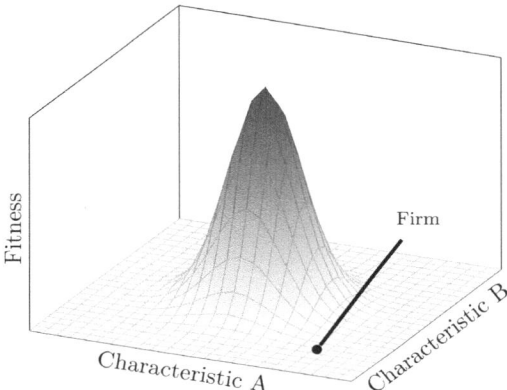

Figure 17: 2-Dimensional Profit Landscape with Homogeneous Demand.

In case of heterogeneous demand, however, consumers and their preferences are also scattered around the (A,B) surface (see Figure 18).

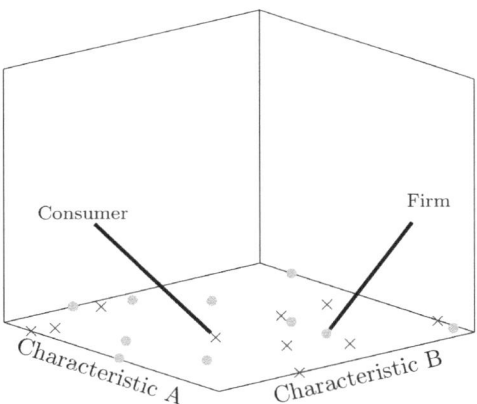

Figure 18: 2-Dimensional Search Problem with Heterogeneous Demand.

As a consequence, no optimal position of firms can be defined as in contrast to the example of homogeneous consumers. Instead of the single peak hill in case of homogeneous demand, the scattered consumer preferences result in a fitness landscape with multiple hills and peaks as shows in Figure 19. As consumers' preferences differ, multiple segments or niches of at least partly similar demand emerge.

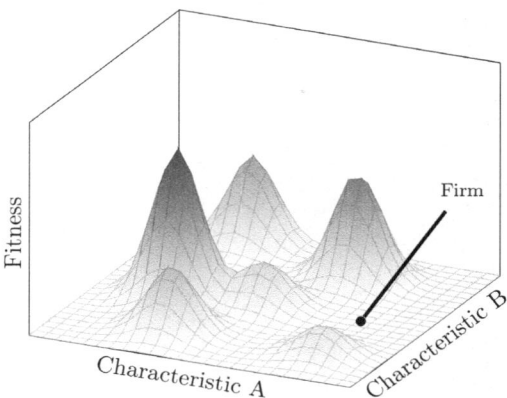

Figure 19: 2-Dimensional Profit Landscape with Heterogeneous Demand.

In the case of heterogeneous demand, the emerging ruggedness in the fitness landscape with its multiple peaks and hills tremendously complicate the search process of firms. This climbing process and the underlying search strategy by firms has been analysed by many scholars. In a single peaked landscape firms are able to simply improve products incrementally to eventually climb the peak of the landscape. In a multi peaked landscape, however, a purely myopic and incremental innovation strategy leads to suboptimal solutions, and as a result other search strategies such as radical changes are of vital importance (Levinthal 1997, Fleming 2001, Fleming, Sorenson 2001).

Although the search problem in the standard NK-Fitness model is of course interesting and yields many implications for innovation science, the focus of the following description of our simulation model lies on the fact that the simulation model in fact represents a dynamic and constantly changing fitness landscape. While the fitness in NK-models usually is fixed, the fitness in our simulation, i.e. the potential sales and profits of a firm's product, depends not only on the demand of consumers but also on the position of products of competitors.

A peak in the fitness landscape represents a position on which a particular firm may find profits by selling its product to a set of consumers. If other firms decide to be active in the market by producing a similar product and hereby *stealing* consumers from the incumbent, the potential profits of all firms change and with them the structure of the fitness landscape.

In other words, in our simulation model the structure of the fitness landscape with its peaks, hills and valleys is not fixed but is the result of the individual position of all firms and consumers. In our baseline simulation model the latter remains fixed over time and, hence, the resulting changes are solely caused by the innovative activities of firms. This changing structure of the fitness landscape caused by the innovative activities of all firms, not only represents a complex search problem for firms, it also induces a constant force for firms to innovate and to hereby adapt their product to the particularities of the whole market (including both consumers' demand and competitors' positions).

To address this important aspect of the simulation it is important to differentiate between the effects of what we call static and dynamic segmentation. With static segmentation we refer to the segmentation of markets induced by the heterogeneous demand of consumers in a static situation, i.e. a situation where the structure of different niches is fixed and solely created by heterogeneous demand. A pure static segmentation would lead to strictly defined niches of consumers with market leaders for every single niche big enough (see for example the models by Chamberlin 1933, Hotelling 1929, Lancaster 1975, 1979, Dixit, Stiglitz 1977).[23]

Dynamic segmentation, in contrast, refers to the process described above where the segmentation of markets is not static but highly dynamic and is driven forth by the innovation activities of firms competing for consumers. In our model, static equilibriums are not achieved. Instead, every innovation by a firm has strong effects on others, causing a chain reaction of new innovations. The changing structure of the theoretical fitness landscape creates a highly dynamic environment for all firms which induces high incentives to innovate.

---

23 Lancaster used his multidimensional description of products and consumer demand also for the analysis of an optimal market structure and market variety based on an equilibrium approach (see for example Lancaster 1975).

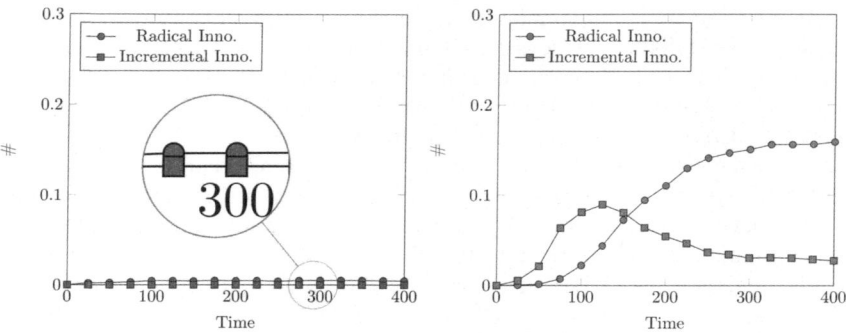

Figure 20: Innovations During a Simulation Run for Homogeneous (left) and Heterogeneous (right) Consumers.

In Figure 20 we present the resulting pattern of the occurrence of radical and incremental innovations by firms during a simulation run for the two extremes homogeneous (left) and heterogeneous (right) consumers. In case of homogeneous demand, we see that innovation activities are limited to radical innovations of a small number of firms. As stated before, in this extreme case the market divides into market leaders and follower firms who do not have any consumers. These followers engage in radical R&D to potentially overtake the market.

In case of heterogeneous demand, only few firms exist in the beginning of a simulation run. These few firms can easily divide the market in segments which explains that the rate of radical and incremental innovation is relatively low. As the market gets more populated by new firms entering the market, both the rate of radical and incremental innovation increases. While during the first 150 time steps incremental innovation dominates, during later stages of the simulation run, the rate of incremental innovation declines. Instead, firms engage more in radical innovation looking for radial new products to find new market niches.

This pattern can be explained by the decrease in technological opportunities. As we assume the size of the market (i.e. the number of consumers and number of possible products) to be fixed, firms face growing difficulties to gain sufficient profits from existing solutions, thus, having to search for new and radical solutions. In the beginning only few firms exist in the market. As a result, every firm can easily gain market shares. As the markets get populated by more firms, the density of market increases and market shares for firms diminish. As a consequence, firms start to engage in incremental R&D. At the later stages of the simulation the market density becomes so high that the small market shares of firms force them to engage in radical R&D. In this situation radical innovation usually creates completely new products which would restart the product-life-circle.

Summing up, we so far analysed with our standard scenario the dynamic effects immediately relevant when allowing for heterogeneous demand. While the situation in a homogenous market with one technological leader seems to be static at least for a short period of time our results show that heterogeneity creates highly unstable dynamics with varying but endogenous incentives to innovate. Instead of creating a situation in which technological leaders for several fixed small niches become established we find a persisting dynamic situation in which firms are unable to occupy stable market shares. Since even for a small niche a full match between products supplied and demanded is highly unlikely, there is always the chance for competitors to entice customers. This, in turn, represents a constant challenge for firms involved in that niche. Firms are forced to engage in R&D activities to regain their market shares which then may entice consumers from other niches forcing other firms to engage in R&D. In other words, if not hindered by other factors and processes such as incomplete information about products or unreasonable loyalty to producers or products heterogeneity on the demand side creates dynamic incentives to innovate endogenously.

4.3.2 Markets in-between Homogeneous and Heterogeneous Demand

To describe the effects introduced by an increased heterogeneity of demand on the innovation processes of firms and the market structure in more detail Figure 21, Figure 22 and Figure 23 depict results showing a stepwise increase of the heterogeneity parameter (p) measured after 500 time steps.

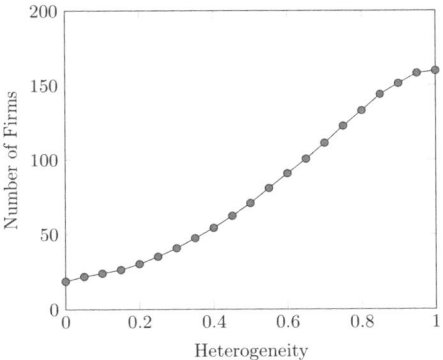

Figure 21: Number of Firms for Different Levels of Heterogeneity.

Our starting point for Figure 21 is again a market in which all consumers share the same demand regarding the product characteristics ($p = 0$). Since only market leaders can successfully sell their products, just a small number of firms survive and the total number of firms in this market is rather small. Increasing heterogeneity of demand (defined by the parameter p) creates a situation where firms sell their product in small niches composed of relatively similar consumers, rather than to all consumers. Through this segmentation more firms can sell their product which leads to an overall increase in the number of firms, forming a s-shaped curve for the total number of firms for increasing heterogeneity.

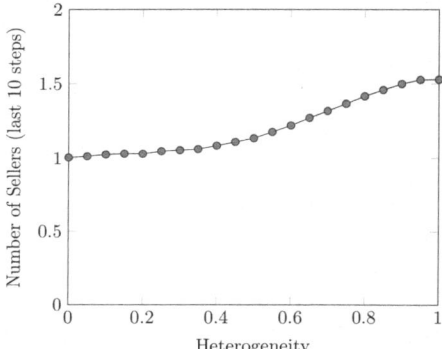

Figure 22: An Indicator of Dynamic Segmentation.

Figure 22 shows the effect of the dynamic segmentation of markets through an increased heterogeneity of demand (see also 4.3.1). To measure the dynamic segmentation, we show the average number of sellers per product category for each consumer over the last 10 time steps of the simulation run. As we can see, for homogeneous markets the situation remains rather static with on average only 1.1 different sellers for each consumer.

Increasing heterogeneity of demand creates niches of similar consumers. However, the drastic raise in the number of firms cannot be explained solely by the existence of different fixed niches. As an additional effect we have to consider the dynamic segmentation explained in subsection 4.3.1. This effect raises the number of different sellers for each consumer over the last 10 time steps by up to 50%, indicating that in this situation firms quickly move through the characteristic space in order to find profitable niches.

Interestingly, this segmentation of the market does not lead to a major change for the average obtained average CPM level (see Figure 23). We expected the average CPM level to extremely drop for increasing heterogeneity as firms' innovation activities have to focus on many different preferences. However, our results indicate that despite firms facing a heterogeneous demand with a broad variety of different preferences in mind by consumers, we find only a (slightly) smaller average CPM level. This leads to the conclusion that the effect of diversified R&D efforts by firms is mostly compensated by the capability of the market to hold firms which engage in R&D (see Figure 21).

Additionally, Figure 23 also contains information about the minimum average achieved CPM level over all 500 simulation runs indicated by the minimum error plots in the figure. On average consumers' preferences are matched with ca. 80% independently on the consumer heterogeneity. In case of heterogeneous demand, we have to account for the fact that some consumers perform better than others. The error plots in Figure 23 shows us the CPM levels of those consumers who perform worst for a given degree of heterogeneity.

As we can see, instead of a strict parallel to the development of the average CPM level in the markets, the minimum CPM levels achieved show their absolute minimum CPM level at a rather low level of heterogeneity. This indicates that especially in case of moderate levels of heterogeneity the risk increases that market fails to match all consumers' preferences sufficiently. However, with increasing heterogeneity this risk decreases and the amplitude of different results decreases.

The reason for that lies in the only moderate increase in the number of firms for low levels of heterogeneity (see Figure 21). Because the distance between consumers increases relatively fast (see Figure 13), for low levels of p, the market does not provide enough firms to cope with the diverse preferences of consumers.

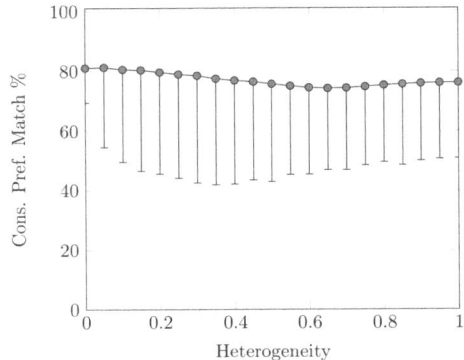

Figure 23: Average CPM Level for Different Levels of Heterogeneity.

The wide range of different CPM levels achieved in case of heterogeneous demand ($p = 0$) is also depicted in Figure 24. While in case of homogeneous demand all consumers share the same preferences and, hence, have by definition identical CPM levels, heterogeneity of demand creates a situation where some consumers' demand is matched better than others. In fact, Figure 24 shows that in case of heterogeneous demand some consumers demand is only matched with 50% while others show CPM levels of over 80%.

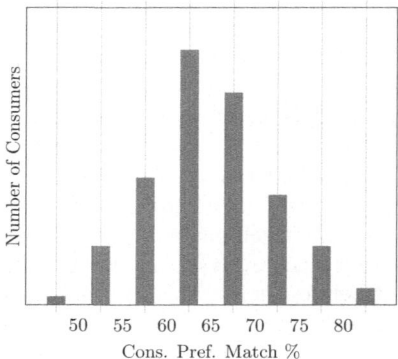

Figure 24: CPM level Distribution.

The market segmentation induced by consumer heterogeneity has also effects on the distribution of sales of firms for different values of p. For investigating the market structure in more detail Figure 25 and Figure 26 provide the average distribution of firm sizes measured through the sales of a firm for different degrees of heterogeneity, measured after 500 steps over 500 simulation runs for all product categories.

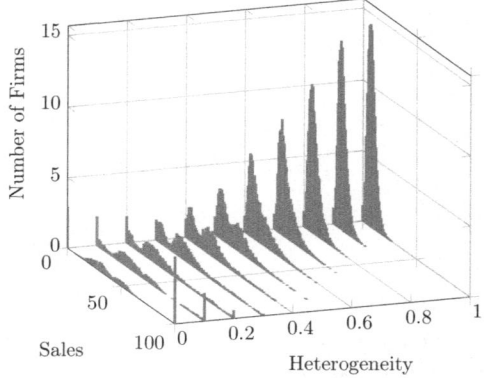

Figure 25: Sales Distribution for Different Levels of Heterogeneity.

While Figure 25 shows the development of the change in sales distributions for ten levels of heterogeneity, Figure 26 shows the respective sales distribution for chosen values of p in more detail. For homogeneous demand ($p = 0$) the results yield a clear division in market leaders (serving 100% of their market niches) and unsuccessful firms with no sales. Additionally, we find a number of firms with sales which cover about 50% of the possible consumers. This occurs because if two firms produce the same product, consumers decide randomly from which one of them to buy, resulting in in a duopoly with two market leader firms.

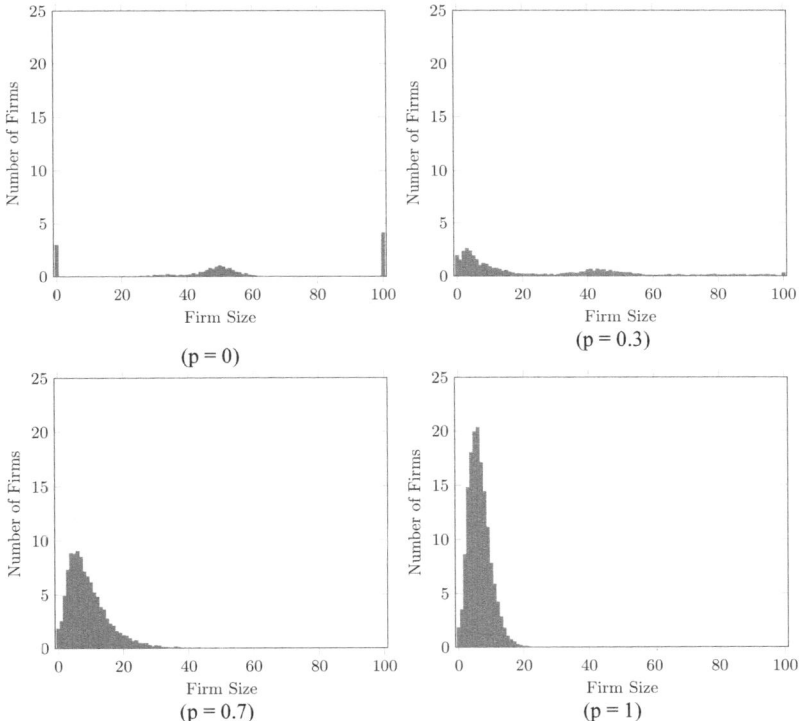

Figure 26: Histogram of Firm Size (for different levels of p).

Increasing heterogeneity creates an entirely different picture. As we can see for $p = 0.3$, already small differences in consumer preferences lead to a more balanced distribution of firms' market shares. Although consumers still share most of their preferences the result is a market with a high number of firms with only few sales and a moderate number of firms with 30% of all sales or more.

Our results also indicate that the heterogeneity of consumer preferences and the resulting segmentation of markets may account for firm size distributions observed by empirical analysis (see for example Gibrat 1931, Lucas Jr 1978, Pavitt et al. 1987). For $p = 0.7$ and $p = 1$ the distribution of sales approximates a right-tailed log-normal distribution for $p = 0.7$ and a standard normal distribution for $p = 1$ as indicated by the literature (see also Mueller et al. 2015 for a more detailed analysis).

As for example Cabral and Mata (2003) point out, the patterns observed by empirical studies are not distinct phenomena but are actually the result of an evolvement over time. While right-tailed log-normal distributions (as generated by our simulation with $p = 0.7$) are generally observed at the time of birth of an industry, the distribution becomes more symmetric over time and approaches a standard normal distribution (as generated by our simulation with $p = 1$). This leads to the potential conclusion that different firm size distributions may actually be explained by a change on the consumer side, i.e. increasing heterogeneity of preferences.

4.3.3   Implications for Innovation Policies

Although heterogeneity of demand can hardly be changed or controlled, the agent-based simulation model still provides useful insights relevant for possible innovation policies. Understanding why markets fail is the basis for sound measures to foster the emergence of innovative markets.

Our results so far show that heterogeneity of demand leads to a dynamic segmentation of consumers, creating a highly dynamic and innovative environment for firms. In contrast, in homogeneous markets only few firms exist. As a consequence, innovation activities are reduced to only some firms engaging in radical R&D in hope to overtake the current market leader. Heterogeneity of demand induces a dynamic segmentation of markets which lead to open markets in which all firms are under constant force to innovate. Thus, as a first application of our simulation model, we stress that these dynamic markets are something desirable. They produce not only tailored products for consumers, but they also create endogenously forces to innovate. Presumably, these innovative markets are also the basis for further developments of new markets and products.

Although the focus of the analysis so far was laid on the heterogeneity parameter p, other parameters of the system, such as the size of the market indicated by the number of consumers or the size of the characteristic space created by the length and number of product characteristics show strong effects on the outcome of the simulation. As our results indicate, the dynamic segmentation of consumers requires both a sufficient number of consumers and products showing a sufficiently large number of distinguishing characteristics for segmentation to occur. For measuring the dynamic segmentation we use the number of firms in the resulting market after 500 time steps in relation to the number of consumers (in this case $p = 1$) and the length of the characteristic space (see Figure 27).

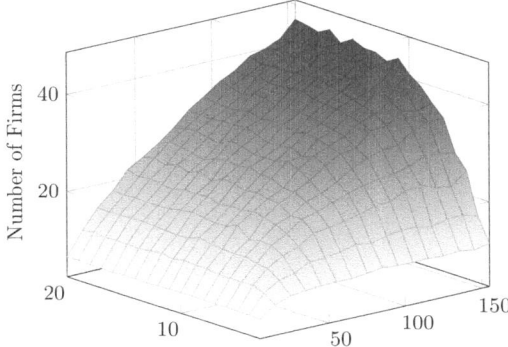

Figure 27: Market Size, Length of Characteristic Space and the Number of Firms.

The characteristic space as expressed by the number of product characteristics and the length of the characteristic bit sequences determines the possible number of segments. In markets where products are less complex, as for example in the market for energy, firms will not be able to differentiate their products from competitors. The same holds true if we consider the limited capabilities of consumers to evaluate product characteristics. Even complex products often are reduced to only few key characteristics. We see this for example for bundled PCs offers where products are advertised by processor speed and the size of the RAM. This limited possibility of firms to differentiate their products, as a consequence, limits the dynamic segmentation and, as we have seen, this finally decreases the innovative efforts of firms.

A sufficient size of the characteristic space however, is not a sufficient condition for market segmentation to occur. While in a homogeneous market the number of consumers is of less importance for the creation of niches, the segmentation of markets induced by the heterogeneity of demand and the complexity of products must be sustained by a sufficient number of consumers. At this occasion, also the role for demand-sided innovation policies designed to create demand becomes relevant. Building on Rogers (2010) and literature studying the diffusion of innovations (see also chapter 2.3) it becomes evident that small number of early adopters of an innovator can be insufficient to create innovative markets. Instead, we need policy measures such as price subsidies to increase the number of early adopters and, thus, foster the emergence of innovative markets.

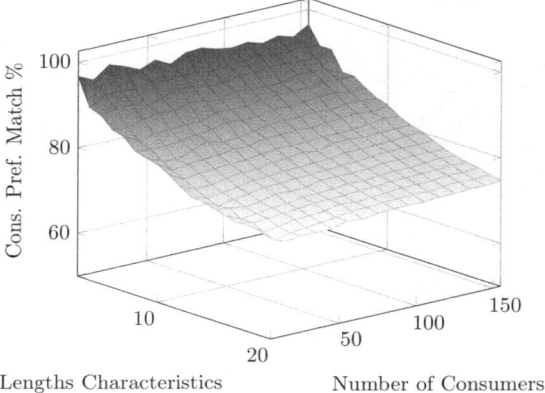

Figure 28: Market Size, Length of Characteristic Space and Average CPM Levels.

A similar result can be seen in Figure 28, where we show the average CPM level obtained after 500 steps for different degrees of product complexity and the number of consumers. Although consumers are of diverse demand, the market only produces products matching the demand if the number of consumers reaches a critical number. Additionally, also the complexity of products in large markets (e.g. with 100 consumers) can to some extend reduce the ability of the market to produce matching products.

Second, we apply our simulation model to illustrate the possibility for policy experiments. So far we have assumed consumers to accept the best offer regardless of the individual match between demand and product. For the following experiments, we implement an additional sensitivity threshold s defining the minimum level of match (i.e. the CPM level necessary to trigger sales). For instance, for $s_n = 70\%$ a product needs to fit at least with 70% of consumers' preferences or consumers will decline the offer i.e. will not buy a product in this period. In order to show the impact of a market with sensitive consumers, Figure 29 depicts the resulting performance of the economy measured in the number of firms after 500 periods for different degrees of consumer sensitivity.[24]

When firms face sensitive consumers with heterogeneous demand for products ($p = 1$) it becomes increasingly problematic to produce products which serve a niche large enough. While this effect seems to be irrelevant for s between

---

24   To show that our starting point usually is a situation where consumers are too sensitive (for high values of s) we reverse the x-axes values in Figure 29.

0% and 60%, the results indicate a major decline in the number of firms in the small area for 60% < s < 80%. Finally, for s > 80% markets collapse. In this situation no firm can survive in the market and as a consequence consumers are not able to buy products.

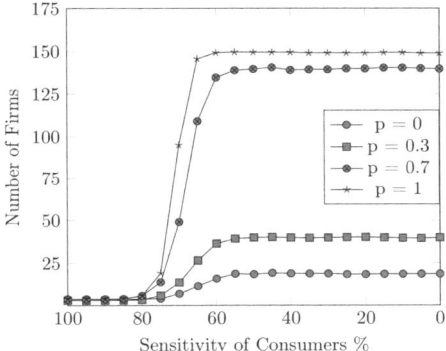

Figure 29: Effect of Sensitive Consumers.

While this effect may be easy to be explained for markets where firms face a heterogeneous demand, it is surprising that this also holds in homogeneous markets. On homogeneous markets the niche size becomes irrelevant because the best products fit every consumer's preferences. In this case, the radical market collapse can be explained by the fact that firms are simply unable to find a composition of knowledge which leads to a product matching the consumers' demand. Because all consumers share the same demand, firms receive no market feedback providing evidence if a firm is roughly matching the demand. Consequently, firms R&D efforts a purely random and destined to fail without further information.

In a final simulation experiment we compare the effectiveness of two different policy instruments. This experiment aims to find out how different policy interventions may foster the creation of a new market. In particular, this experiment is suited to analyse the different impacts of policy instruments targeting either the supply or the demand side. Within our simplified model, one possible effect of supply side subsidies can be seen as switching from a high R&D threshold to a low R&D threshold. Subsidies in this sense financially support firms in the market and reduce the minimum number of consumers necessary, below which firms try to change their product by innovation. Contrastingly, demand-oriented policies designed to trigger development in a new industry can directly target consumers' decisions through price subsidies. Making new product cheaper will decrease the sensitivity of consumers and reduce the minimum level of compatibility between demand and supply.

In the following experiment we compare the development of the number of firms for different subsidies levels and targets. Figure 30 shows for both R&D strategies (red and blue plot in Figure 30) that, given a high sensitivity (s_n) of consumers, functional markets fail to emerge. For s between *55%* and *80%* the effect of a lower threshold for research, measured as the difference between both curves (i.e. the two R&D strategies) is positive. For s_n < *55%* the effect reverses and a higher R&D threshold creates an economy with a higher number of firms.

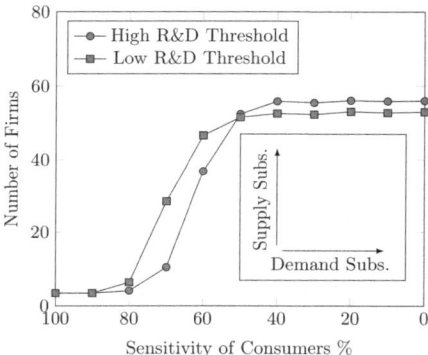

Figure 30: Number of Firms for Different Policy Settings.

As stated before, the reason for the economy to fail for a high s for both R&D thresholds lies in the fact that firms find themselves unable occupying niches containing a satisfactory number of consumers. This can be explained by (a) the firms' expectations about the niche size are too high or (b) the high level of s hinders the pooling of consumers to a niche where one product can satisfy the preferences of different consumers.

Although both parameters (*R&D threshold* and *consumers' sensitivity*) represent abstract concepts, they are suited for an interpretation from a short-term policy perspective. As discussed above, one possible policy measure to influence the required match between demand and supply can be found in subsidies on the consumer side. Secondly, subsidies can also be seen to affect firms' R&D strategies if, for example, the research and production costs of a firm are subsidised. This decreases the minimum market share, which is considered to be necessary. In the same vein, one can interpret the results in Figure 30 as follows: starting at a point where our simulated economy fails ($s = 100\%$), firm-oriented subsidies affecting the R&D strategy do not have an effect on the emergence of the industry as for both R&D strategies the number of firms stays low. Instead, the markets need to be subsidised via the demand side and, thus, reducing the sensitivity of consumers.

Actually, only within the small range of s between 80% and 55% subsidies for firms can have a positive effect which is indicated through the increase in the number of firms. However, the results do not indicate whether these subsidies are more efficient than demand-side subsidies, which also have a positive impact. Finally, this positive effect on the number of firms reverses again with smaller compatibility rates ($s_n < 55\%$).

In other words, our results indicate that if consumer preferences are heterogeneous and the sensitivity of consumers is high, it is impossible for firms to match the preferences of enough consumers to create a niche of sufficient size. In this situation, the R&D threshold would have to be decreased to a point where already a niche of one consumer would be acceptable for the firm to stay in the market. In contrast to this, the results show that subsidising the consumer side and decreasing the sensitivity of consumers opens up the market so that small niches can emerge in the first place.

4.4 Discussion

In this chapter we introduced an agent-based model of innovation, knowledge creation and heterogeneous demand. Innovative firms try to adapt their knowledge stock through R&D to fit the individual preferences of consumers. In contrast to other approaches, our model follows Lancaster (1966) and incorporates heterogeneity of demand as the individual preferences for product characteristics such as colour, size or functionality. The aim, however, is not to display the economy with a high degree of accuracy, instead, our aim is to understand the fundamental changes if we consider heterogeneity of demand.

Consequently, the first challenge was to find an adequate representation of knowledge and a suitable and flexible method to transform knowledge into marketable products and the corresponding environment. With our approach of a knowledge representation as bit sequences and with the use of mapping functions we can implement a coding of knowledge to products which represents *substitution, interdependence, transparency* which is *variable* and *computational*.

While the processes introduced may be considered simplistic they are nonetheless able to reproduce fundamental aspects, which have to be considered for the analysis of innovation processes. The first striking result is that consumer heterogeneity itself creates static and dynamic segmentation of markets and by that a persistent innovation dynamic beyond any equilibrium consideration without reaching a steady state. Already a small deviation from perfect homogeneous demand creates niches, which open the market for a high number of firms. Firms find themselves in a highly competitive situation, which creates constant and endogenous incentives for innovation and knowledge creation. In this situation every temporary stable market can be easily destroyed by the innovative activities of competitors or market entrants. At this occasion it is also important to emphasize on the differences of our approach to the analysis of optimal product variety studied for example by Harold Hotelling (1929) and Kevin Lancaster (1966, 1975, 1979). As our model shows, heterogeneous demand itself does not create stable equilibria. In fact, homogeneous markets lead to at least temporally monopolistic outcomes and tend to reduce the innovative power of markets.

Second, we found that the simulation model is capable of showing that log-normal as well as normal firm size distributions can be explained by the heterogeneity of consumers. Since the findings of Robert Gibrat (1931) several authors have looked at the firm size distribution (FSD) and received widely shared stylized fact is that the FSD is stable and approximately log-normal (Cabral, Mata 2003). Recent empirical evidence shows that the FSD in more complete data sets may evolve over time and differ from a log-normal distribution (Evans 1987, Hall 1987, Cabral, Mata 2003). In this context our model places the heterogeneity of demand into perspective leading to the hypothesis that a possible explanation for this is an increasing level of heterogeneity and differentiation in the respective markets.

Finally, our simulation environment allows for experiments relevant also for policy measures. As a result of this experiment, we claim that a sound design of innovation policies to foster industry development has to consider both sides of the market and to carefully evaluate which market side should be influenced by policy interventions. In more detail, our simulation model shows that a dynamic and innovative market design only occurs if heterogeneous demand, complexity of products and a sufficient number of consumers come together. Additionally, the results indicate that the effectiveness of demand side and supply side subsidies depends strongly on the current situation of the market and that supply side subsidies are only suitable under certain circumstances and may eventually cause major drawbacks.

Obviously, the simulation model allows for wide range of extensions and applications. Although introducing heterogeneity of demand is without doubt an important step for the analysis of innovation processes, consumers are still represented inadequately if considered in isolation, neglecting the vast forms of direct or indirect interaction between them (see for example von Hippel 1988, Von Hippel 2005, Valente 2009). The following chapter 5, therefore, introduces additional extension to the baseline model of this chapter.

5 Networks of Heterogeneous Agents

Introducing heterogeneous demand of consumers presents without doubt a huge step towards a more realistic consideration of consumer into the analysis of innovation processes. However, the baseline model shown in chapter 4 still over simplifies the capabilities and the behaviour of consumers in various ways. Following the idea of *population thinking* (see also chapter 2.2) we have to look at the social and economic interactions within the populations of heterogeneous actors. However, we have to consider that agents in the baseline model (firms with firms and consumers with consumers) only interact indirectly with each other. Consequently, this chapter aims to extend the baseline model by introducing a network perspective.

We start in section 5.1 with a theoretical extension of the baseline model in chapter 4 in which informal firm networks are introduced. Building on the findings of section 5.1, section 5.2 implements boundedly rational consumers and networks between them.

5.1 Informal Knowledge Exchange in Firm Networks

In this section we apply an agent-based simulation approach to explore how and why typical network characteristics affect overall knowledge diffusion properties between firms. After some introducing remarks in section 5.1.1, in subsection 5.1.2 we describe related research, the barter trade diffusion process used for the diffusion of knowledge, and finally the four network topologies on which our analysis builds on. In subsection 5.1.3, we present a simulation model designed to analysis in detail the outcomes of different network characteristics for the diffusion of knowledge. The results are finally discussed in section 5.3 together with some remarks on limitations and fruitful avenues for further research.

5.1.1 Introducing Remarks

The method behind the study of networks in social science, today commonly known as *Social Network Analysis* (SNA) can be traced back to the work of authors such as Jacob L. Moreno and Helen Jennings who began in the early 20th century to systematically push forward this method as the study of social interactions between individuals (Freeman 2004, Borgatti et al. 2009). Today, it has gained wide acceptance among social sciences, physics, biology and many other scientific fields and for example is also used to model the interactions between firms.

In this chapter we apply a structural perspective on networks. Economic growth and prosperity is, without doubt, closely related to innovation processes which are fuelled by the ability of the actors involved to access, apply, recombine and generate knowledge. Consequently, the term *knowledge-based economy* has become a catch phrase. Knowledge-based econom*ies are* "directly based on the production, distribution and use of knowledge" (OECD 1996, p. 7). This recognition led to a growing interest in knowledge generation and diffusion among practitioners, politicians and scholars alike.

It is important to note that information and knowledge is not freely available or homogenously distributed among the actors of a real-world economy. The simplifying neoclassical notion of knowledge as a ubiquitous public good does not reflect everyday reality. Instead, innovators are constantly searching for new ideas, opportunities, and markets. The neo-Schumpeterian approach to economics (Hanusch, Pyka 2007c) emphasises the role of innovators and explicitly acknowledges the nature of information and knowledge. Malerba (2007, p. 16) argues that knowledge and learning are key building blocks of the neo-Schumpeterian approach. At the same time, this approach accounts for the firms' ability to store and generate new stocks of knowledge by referring to the concept of organisational routines, originally developed by Nelson and Winter (1982).

The traditional idea that knowledge can be acquired without any restrictions and obstacles has been replaced by other, more realistic concepts and explanatory approaches. According to Malerba (1992), firms can gain access to new knowledge via internal processes (e.g. intra-organisational learning) or by external knowledge channels (e.g. modes of inter-organisational cooperation). We particularly focus on the latter channel. It has been argued that networks provide the pipes and prisms of markets (Podolny 2001) while enabling the flow of information and knowledge. The ties between the nodes in such networks can be both formal and informal (Pyka 1997, p. 210).This chapter particularly focuses on firms' external knowledge channels and analyses informal cooperation and network structures.

In line with our previous considerations, Herstad et al. (2013, p. 495) point to the fact that "[...] innovation is shifting away from individual firms towards territorial economies and the distributed networks by which they are linked". Inter-organisational knowledge exchange is of growing importance for the competitiveness of firms and sectors such that innovation processes nowadays take place in complex innovation networks in which actors with diverse capabilities create and exchange knowledge (Levén et al. 2014, p. 156). Networks are the prerequisite for the exchange of knowledge. As these systems are becoming more and more complex, the linkage between the underlying network structure and knowledge creation and diffusion have to be analysed and understood thoroughly.

The natural question that arises in this context is what do we know from previous research on the issues raised above. To start with, several studies have focused on the efficiency of knowledge transfer in networks with either regular, random or small world structures (Cowan, Jonard 2004, Morone et al. 2007, Lin, Li 2010). So for example, Cowan and Jonard (2004) use a barter trade diffusion process to investigate the efficiency of different network structures. Their model shows that, unlike fully regular or random structures, small-world structures lead to the most efficient (as well as the most unequal) knowledge diffusion within informal networks. Building on these findings, Morone et al. (2007) use a simulation model to analyse the effects of different learning strategies of agents, network topologies, and the geographical distribution of agents and their relative initial levels of knowledge. The results show that, in contrast to the conclusions drawn by Cowan and Jonard (2004), small-world networks do perform better than regular networks, but consistently underperform compared with random networks.

A new aspect in this debate was introduced by Cowan and Jonard (2007). The study is based on the theoretical discussion on structural holes (Burt 1995) and social capital (Coleman 1988). Cowan and Jonard introduce a simulation model with a barter trade knowledge exchange process in which the authors analyse the effects of network randomness and the existence of stars (i.e. firms with a high number of links) on a systemic and individual level. Their results show that the existence of stars can either be good or bad for the diffusion of knowledge depending on whether stars are givers or traders of knowledge.

The aspect of different degree distributions of networks is also stressed by Lin and Li (2010). The authors analysed how different network structure affect knowledge diffusion in a scenario in which agents freely give away knowledge. Their analysis reveals that networks with an asymmetric degree distribution, namely scale-free networks, provide optimal patterns for knowledge transfer.

The aim of this chapter is twofold. On the one hand, we conduct several simulation experiments to gain an in-depth understanding of how network characteristics, such as path length, cliquishness and the distribution of degrees, affect the knowledge distribution properties of the system. We study the interplay between these network characteristics to gain an in-depth understanding how mutually interdependent processes affect the diffusion of knowledge among the actors involved. On the other hand, the simulation model allows also for *in-silicio* investigating the implications of the findings through a virtual policy laboratory. In this policy laboratory, we can systematically change and alter the network structure, analysing the outcome of different measures designed to foster knowledge diffusion.

With this model we analyse how the structural properties of four structurally different network topologies affect the overall knowledge diffusion properties within these networks. To do so, we implement a simple barter trade knowledge diffusion process in an agent-based simulation model which presents a theoretical extension of the simulation model introduced in chapter 4. However, given the narrow focus of this chapter's analysis we exclude the market processes shown in chapter 4 and consider only firms (i.e. agents) interconnected within a static network of a predefined size and density without any market processes. This allows us to avoid unintended interferences between market processes, and the diffusion of knowledge.

5.1.2   Knowledge Exchange and Network Formation Mechanisms

This section explains the modelling background of the simulation analysis in section 5.1.3. We start this section by explaining the knowledge barter trade diffusion model. Subsection 5.1.2.2 then describes the key characteristics of the four different network algorithms used for the analysis.

5.1.2.1   The Barter Trade Process

For our simulation model we use a barter trade diffusion model representing informal knowledge exchange first introduced by Cowan and Jonard (2004). This barter trade knowledge diffusion process is modelled as follows:

Every agent $i \in I$ is endowed with a knowledge vector $k_i = (k_{i,c})$ with $i = 1, ..., I; c = 1, ..., K$ different knowledge categories. Knowledge is exchanged between agents in a barter exchange process in a sense that they trade knowledge if trading is mutually beneficial. An exchange therefore takes places if two agents are directly connected via a link and if both agents can receive unknown knowledge from the respective other agent, independent of the amount of knowledge they actually receive. This assumption allows us to incorporate the realistic idea that agents can only assess whether or not the potential partner has some relevant knowledge to share and not to a priori assess how much can be gained exactly from the knowledge exchange. This is in line with the particularity of knowledge that its exact value can only be assessed after its consumption (if at all).

In a more formal description, two conditions have to be fulfilled. Let $j \in N_i$ and assume there is a number of knowledge categories $n(i,j) = \#\{c: k_{i,c} > k_{j,c}\}$ in which agent i's knowledge strictly dominates agent j's knowledge. As we already know, agent j will only be interested in a trade with agent i if $n(i,j) > 0$ and *vice versa*. Hence, the barter exchange takes place and agents i and j exchange knowledge if and only if first, $j \in N_i$, and if second, $min\{n(i,j), n(j,i)\} > 0$. This is also called a *double coincidence of wants* (Cowan, Jonard 2004, p. 1562). If the double coincidence of wants condition holds true, the agents exchange knowledge in as many categories of their knowledge vector as mutually beneficial. If the number of categories in which the agents strictly dominate each other is not equal among the trading agents (i.e. $n(i,j) \neq n(j,i)$), the number of categories in which the agents exchange knowledge will be equal to $min\{n(i,j), n(j,i)\}$, while the decision in which categories the agents eventually exchange knowledge is randomly chosen with a uniform probability. Besides the particularity of knowledge exchange presented above, the model also incorporates the fact that the internalization of knowledge is difficult and the exchange of knowledge is only partly possible due to the different absorptive capacities of the agents. This means that only a constant share of α with $0 < \alpha < 1$ can be actually assimilated by the receiver. Therefore, each period in time the knowledge stock of an agent can either increase to a before the exchange unknown amount (if an exchange takes place) or stay constant (if no exchange takes place).

Agents in the model mutually learn from each other and by doing so knowledge diffuses through the network and the mean knowledge stock of all agents within the network $\bar{K} = \sum_{i \in I} K_i / I$ increases over time. As knowledge is considered to be non-rival in consumption, the knowledge stock in the economy can only increase or stay constant, but an agent will never lose knowledge by sharing it with other agents. Assume, for instance, that $n(i,j) = n(j,i) = 1$ and that in category c_1 agent j's knowledge strictly dominates agent i's knowledge and that in category c_2 agent i's knowledge strictly dominates agent j's knowledge. In this situation agent i will receive knowledge from agent j in category c_1 (with his knowledge in category c_2 being unaffected) and agent j will receive knowledge from agent i in category c_2 (with his knowledge in category c_1 being unaffected). Therefore, after the trade the knowledge of agent i changes according to $k_{i,c1}(t + 1) = k_{i,c1}(t) + \alpha(k_{j,c1}(t) - k_{i,c1}(t))$ and the knowledge of agent j changes according to $k_{j,c2}(t + 1) = k_{j,c2}(t) + \alpha(k_{i,c2}(t) - k_{j,c2}(t))$. As agents exchange their knowledge as long as this trade is mutually advantageous, the barter trade process takes place until all trading possibilities are exhausted, i.e. there are no further double coincidences of wants.

5.1.2.2 Algorithms for the Creation of Networks

To investigate the structural effects of different network topologies on knowledge diffusion we apply four structurally distinct algorithms to construct network topologies. The four resulting network topologies are the *Erdös-Rényi* or *random* network (ER), the *Barabási-Albert* network (BA), the *Watts-Strogatz* network (WS), the *Evolutionary* network (EV). While ER, BA, and WS networks are well-known, EV networks are considered because of their realistic network formation strategy and because of their unique network characteristics.

We begin by briefly addressing ER networks (Erdős, Rényi 1959, 1960). The attachment logic behind the *Erdös-Rényi* (n, M) algorithm is quite simple. Each of the n nodes attracts ties with the same probability p which ultimately creates M randomly distributed links between the nodes. The resulting random graphs are characterised by short path length, low cliquishness and a relatively asymmetric degree distribution following a Poisson or normal degree distribution (Erdős, Rényi 1960, Bollobás et al. 2001).

Barabási and Albert (1999) introduced a preferential attachment mechanism which better explains structures of real-word networks compared to random graphs. The degree distribution of links among nodes approximately approaches a right skewed power law where a large number of nodes have only a few links and a small number of nodes are characterised by a large number of links. This is usually described by the following expression: $P(k){\sim}k^{-\gamma}$, in which the probability $P(k)$ that a node in the network is linked with k other nodes decreases according to a power law. The BA algorithm is based on the following logic: Nodes with above average degree attract links at a higher rate than nodes with fewer links. More precisely, the algorithm starts with a set of 3 connected nodes. New nodes are added to the network one at a time. Each new node is then connected to the existing nodes with a probability that is proportional to the number of links that the existing nodes already have. This process of growth and preferential attachment leads to networks which are characterised by small path length, medium cliquishness, highly dispersed degree distributions - which approximately follows a power law -, and the emergence of highly connected hubs.

Watts and Strogatz (1998) stressed that biological, technical and social networks are typically neither fully regular nor fully random but exhibit a structure that is somewhere in between. In their seminal study they proposed a simple algorithm which enabled them to reproduce so called small-world networks. The algorithm defines the randomness of a network using the parameter $w \in [0, ... ,1]$ that describes the probability that links within a regular ring network lattice are redistributed randomly. The authors found that, within a certain range of randomness, the resulting networks show both a high tendency for clustering, like a regular network, and at the same time, short average paths lengths, like in a random graph.[25]

Finally, our last network algorithm is the EV algorithm originally proposed by (Mueller et al. 2014). The algorithm was developed to create dynamic networks in a well-defined population of firms. We employ the EV algorithm to study

---

25   The exact rewiring procedure works as follows: The starting point is a ring lattice with n nodes
      and k links. In a second step, each link is then rewired randomly with the probability w. By
      altering the parameter w between $w = 0$ and $w = 1$, i.e. the network can be transformed from
      regularity to disorder.

knowledge diffusion processes[26] for at least two reasons: it is based on a quite realistic linking strategy of the actors at the micro level and it allows network topologies to be generated which combine the characteristics of WS and BA networks. The underlying idea of the algorithm is straightforward. The EV algorithm is based on the notion that innovating actors typically face a scarcity of information about potential cooperation partners. As a consequence, actors are continuously adapting their partner selection behaviour to address this information deficit problem. The trade-off between the need for reliable information and the cost of the search process is reflected in a two-stage selection process in which each node chooses link partners based on both the transitive closure mechanism and preferential attachment aspects. For every time step, each agent defines a pre-selected group consisting of potential partners which they know via existing links, i.e. cooperation partners of their cooperation partners for which no link currently exists. If the size of this pre-selected group is smaller than a defined threshold, agents are added randomly to create a pre-selection of defined size. In a second step, agents then choose the potential partner with the highest degree centrality and form a link.

5.1.3   Model Analysis

We present the findings of our simulation analyses where we explore how different network topologies affect the diffusion of knowledge. First, we address the effect of network characteristics, such as path length and cliquishness, on network performance in terms of the average knowledge level of all actors in the network (subsection 5.1.3.1). We then investigate how the distribution of links among these actors affects network performance (subsection 5.1.3.2). Finally, we run policy experiments for each of the four networks to gain an in-depth understanding of how policy interventions may affect the diffusion of knowledge (subsection 5.1.3.3).

5.1.3.1   Path length, Cliquishness and Network Performance

The model is initialised with a standard set of parameters as follows: we assume a model population of $I = 100$ firms connected by 200 links for all networks. The agents and links within the network are placed according to the algorithm described before which leads to the following four networks: (i) Random - Erdös / Renyi (n,M), (ii) Watts-Strogatz, (iii) Barabási-Albert and (iv) Evolutionary

---

26   In this paper we analyse diffusion processes in existing networks. In the case of the EV algorithm we assume that the linking process is repeated 100 times. To create comparable networks with a pre-defined number of links we further assume that links are deleted after 2 time steps of the rewiring process.

network algorithm. More precisely, we assume for the Watts-Strogatz algorithm a probability $w = 0.15$ and for the Evolutionary network algorithm a preselection group of 5 and 100 time steps for the initiation of the network. Figure 31 illustrates the network patterns produced by these formation algorithms.

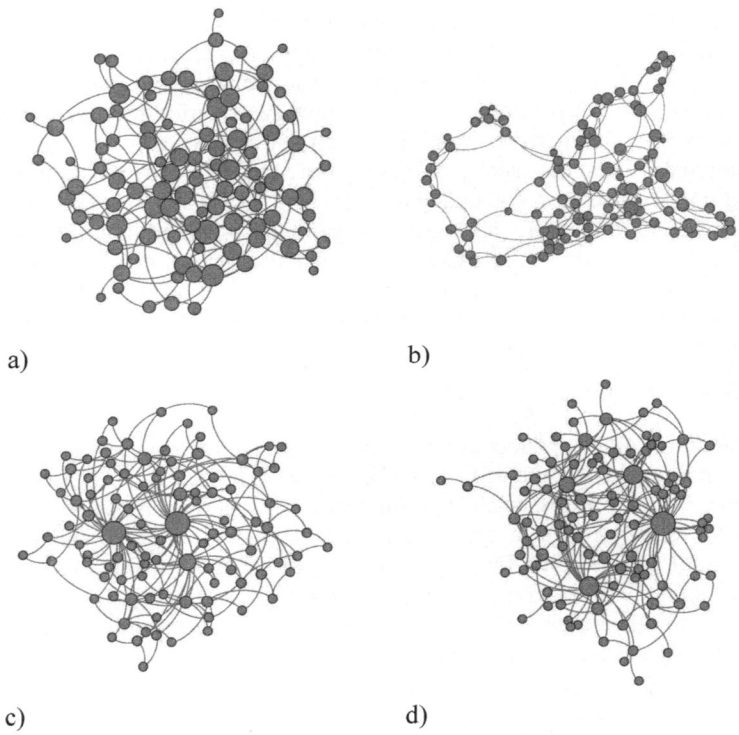

a)   b)

c)   d)

Figure 31: Networks within the Simulation Model; a) Random - Erdös / Renyi (n,M) b) Watts-Strogatz c) Barabasi-Albert d) Evolutionary Network Algorithm.

Following Cowan and Jonard (2004), for each model run, each agent is equipped with a knowledge vector $k_{i,c}$ with 10 different knowledge categories drawn from a uniform distribution, i.e. $k_{i,c}(0) \sim U[0,10]$. To enhance the knowledge diffusion, we also define 10 randomly chosen agents as 'experts', i.e. these agents are endowed with a knowledge level of 30 in one category. Unless stated otherwise, we assume a value of $\alpha_i = 10\%$ for the absorptive capacities. Finally, we assume that the knowledge levels of the agents in one category is similar if the difference in this respective category is less than 1%.

Figure 32 shows the average overall knowledge stock in the four networks over time, i.e. the mean average knowledge $\bar{k}_t = \sum_{i \in I} k_{i,t} / I$ of all agents within the network averaged over 500 simulation runs as well as the error bars of the respective results. The black part of the plot indicates that the knowledge stock in the network is still growing (knowledge exchange is still taking place), while the grey part of the plot indicates that the knowledge stock is no longer changing (knowledge stock has reached a steady state). The figure shows that the average knowledge stocks within the networks increase over time, however, there are significant differences between the four network topologies: WS networks perform best followed by ER networks, BA networks, and EV networks. It can also be seen that in the worse performing networks, the maximum knowledge stock is reached earlier than in the better performing networks. In the two best performing networks, knowledge is diffused for nearly twice as long as in the worst performing evolutionary networks.

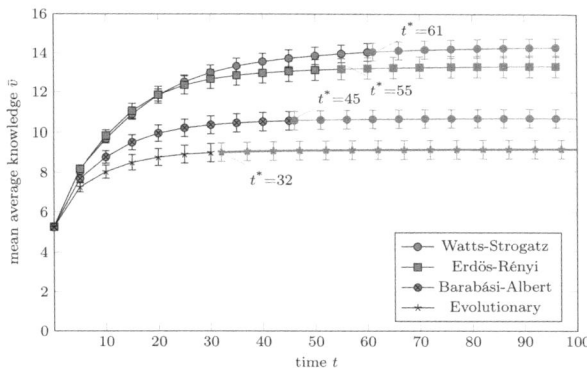

Figure 32: Average Knowledge Levels of Agents.

In Table 3, we present network characteristics, i.e. average path length and global cliquishness of our four network topologies. Following the idea that path length and cliquishness are the main factors influencing the diffusion of knowledge, we now can investigate the relationship between these two characteristics and the diffusion performance shown in Figure 32.

Our results reveal a positive relationship between path length and the average knowledge levels of nodes for the four network topologies. In fact, the networks with the lowest average path length are the networks that perform worst, i.e. the EV networks. However, the second network characteristic shown in Table 3 also fails to coherently explain the simulation results. While WS networks, which show the highest level of cliquishness, also result in the highest average knowledge levels, networks with the second best performance (ER networks) have the lowest average cliquishness. Moreover, the average path length of the BA and EV networks are quite similar, however, the average cliquishness of EV networks is twice the cliquishness of BA networks and yet BA networks still outperform EV networks.

Table 3:   Average Path Length and Cliquishness of all Four Network Topologies.

	Watts-Strogatz	Erdös-Rényi	Barabási-Albert	Evolutionary
Path length	4.49	3.45	2.99	2.74
Clustering Coefficient	0.32	0.03	0.13	0.27

Interestingly, these patterns hold true even for different values of absorptive capacities (see Figure 33). To analyse the effect of different values of absorptive capacities Figure 33 depicts the average knowledge levels in the networks for different values of α. To be able to compare the results we extend the number of steps analysed to 1000 which ensures that in all cases the diffusion process stop and reach a final state.

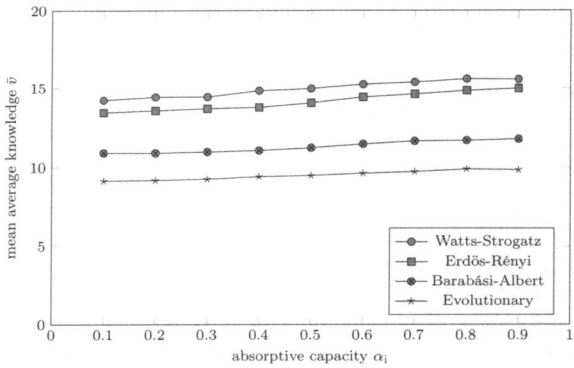

Figure 33: Knowledge Levels of Agents for Different Absorptive Capacities.

These counter-intuitive results prompt the question of whether a network's path length and its cliquishness can fully explain the performance differences between the observed networks. Following the idea of Cowan and Jonard (2007), another network characteristic that may explain the differences in network performance is the distribution of links among agents.

The authors found that, in a barter economy, network performance is negatively affected by the existence of stars with a relatively high degree centrality. According to the authors, this is because stars have so many partners that they acquire all the knowledge they can in a very short time. This rapidly leads to a lack of double coincidences of wants which stops the knowledge trading process within the network, and which, in turn, may even disconnect the entire network (Cowan, Jonard 2007, p. 108):

> "If the stars are traders, because they have many partners, they will rapidly acquire all the knowledge they need, and so stop trading. This blocks many paths between agents, and […] can disconnect the network."

To test this hypothesis, we conduct in the following subsection several simulation experiments to show to what extend the degree distribution of a network has an effect on the diffusion processes and if the provided explanation fully explains the obtained differences in the simulation results.

5.1.3.2   Degree Distribution and Network Performance

Next, we explore the relationship between degree distribution and network performance. The information depicted in Figure 34 shows that the networks analysed in this paper do not only differ significantly in terms of their performance, but also with respect to the distribution of degrees. The worst performing networks, i.e. BA and EV networks, are networks that have a highly skewed and dispersed degree distributions which approximately follow a power law. Even though all networks by definition have the same average degree of 4, BA and EV networks are characterised by a large number of small nodes (having only a few links) and a few nodes with an extremely high degree. In contrast, WS and ER networks have more symmetric degree distributions with only small deviations from the average degree of the network. These better performing networks are characterised by a less asymmetric degree distribution. So, by only looking at the degree distributions, the worst performing networks indeed have a more asymmetric degree distribution than the better performing networks.

To further analyse the effect of the degree distribution on diffusion performance, we show in Figure 35 the relationship between the variance of the degree distribution and the mean average knowledge level in the respective networks achieved after 100 simulation steps. In contrast to path length and cliquishness, the variance of the degree distribution of networks has a coherent effect on network performance. WS networks, which perform best, are characterized by the lowest variance in the nodes' degrees. This inverse relationship between variance and network performance also applies to the other networks. For example, the worst performing networks, EV networks, are also those networks with the highest variance in their degree distribution.

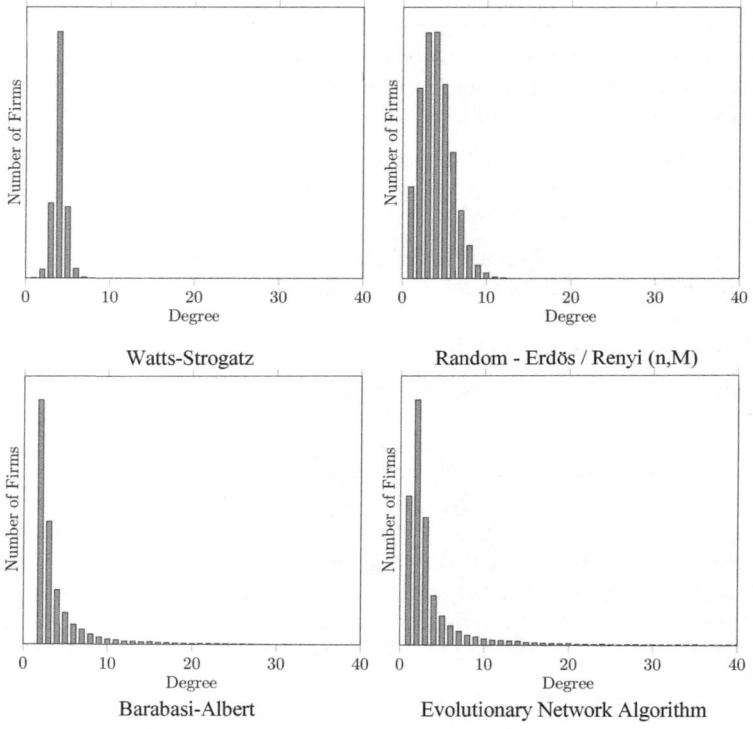

Figure 34: Degree Distribution in the Four Networks.

However, in contrast to the explanation by Cowan and Jonard (2007), our results indicate that the weak performance of networks with a highly skewed degree distribution cannot (exclusively) be explained by stars which stop the barter trade process and eventually disconnect the network. Instead, our results indicate that the dissimilarity between nodes and the resulting existence of relatively small, inadequately embedded nodes actually stops the diffusion of knowledge within networks.

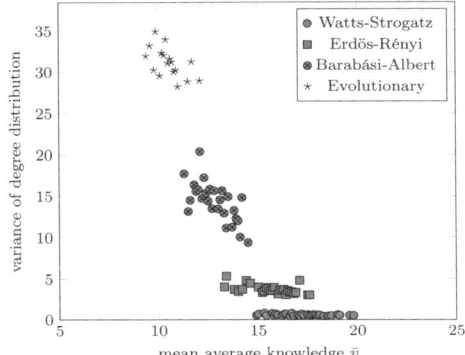

Figure 35: Variance of Degree Distribution and Average Knowledge Levels.

Figure 36 illustrates the cumulative number of agents that stopped trading over time. Not surprisingly, in the worse performing networks, agents stop trading earlier than in the better performing networks. Comparing EV and WS networks after 40 time steps showed that almost 90% of the agents in evolutionary networks had already stopped trading, whereas 65% of the agents were still trading in WS networks. Moreover, in EV networks almost all agents stopped trading after 70 time periods, whereas in WS networks this only occurred after 100 time periods.

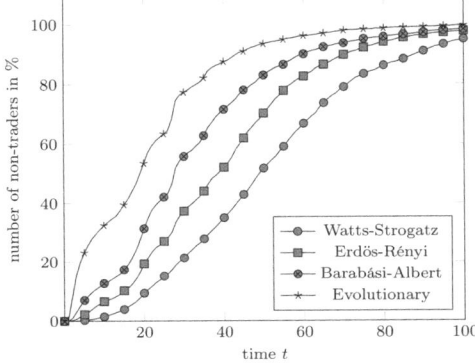

Figure 36: Cumulative Number of Non-Traders.

The results gained so far, indeed provides evidence that the reason for the poor performance of networks with asymmetric degree distribution are agents who stop trading early and, hence, disrupt the knowledge flow. However, the question at hand is: why?

To answer this question, Figure 37 and Figure 38 show the relationship between the nodes' degrees and the time when these nodes stop trading as well as the relationship between the nodes' degrees and their knowledge level acquired over an average over 500 simulation runs.

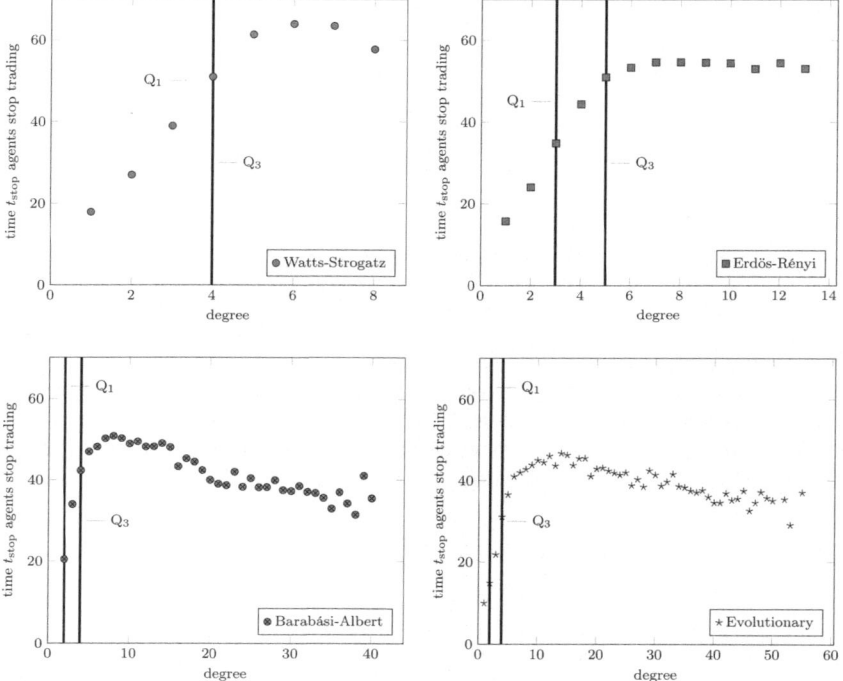

Figure 37: Relationship Between the Time Agents Stop Trading and Their Degree.

Figure 37 shows that in general, we find a positive relationship between the size of an agent and the time it stops trading, especially for networks with dispersed degree distribution (i.e. BA and EV networks). This positive relationship always holds true for the first three quartiles (the lower 75%) of the distribution. In the lower 75% of the distribution, nodes actually stop trading earlier the fewer links they have. However, we see that, especially in networks with asymmetric degree distribution, this relationship fails for nodes with an extremely high number of links (the fourth quartile or the upper 25% of the distribution).

From this exploration we can conclude that nodes with a high number of links do not stop trading first. In contrast, our results indicate that small nodes (the lower two quartiles of the distribution) stop trading first. Big nodes stop trading later, however, medium sized nodes trade the longest. As a consequence, the poor performance of networks with a dispersed degree distribution can be explained by the sheer number of small nodes. Interestingly, as indicated by the quartiles of the BA and EV networks, it is important to note that when we speak of small nodes that stop trading earlier than big nodes, we are referring to the majority of nodes, i.e. over 50% of all nodes. If we now combine the information values provided by Figure 37 and Figure 32, we also see that nodes with a high degree actually stop trading after the increase in knowledge levels has reached its peak and almost no knowledge is traded within the network anymore.[27]

To further investigate why nodes stop trading, we illustrate, in Figure 38, the relationship between node's degree and the mean knowledge these nodes reach after 100 time steps averaged over 500 simulation runs. Additionally, the size and the colour of the marks indicate the reason why nodes do not trade knowledge. The data clearly shows a positive relationship between a node's degree and its acquired knowledge level. So, in general, the more links a node has, the more knowledge it will receive. This positive effect, however, decreases for nodes with many links, indicating a saturation phenomenon for nodes with a high number of links in the network.

In additional to the relationship between degree and the time nodes stop trading, Figure 38 also shows us why these nodes eventually stop exchanging knowledge.[28] While white marks indicate that nodes with the respective degree centrality stop because they could not offer knowledge to their trading partners, black marks indicate that these nodes stop trading because their partners do not offer them enough knowledge as a sufficient barter object. As the figure shows, especially small nodes stop trading because they cannot offer knowledge to their partners. Nodes with a high degree stop trading because their partners cannot offer new knowledge. Medium-sized nodes (with grey marks), by contrast, stop trading because they have too much knowledge for their small partners and too little knowledge for their very large partners.

Summing up, we can confirm the results by Cowan and Jonard (2007) that for barter trade diffusion processes the degree distribution of nodes is of decisive importance, even more important than other network characteristics such as path length and cliquishness. Based on our simulation results, we come to the

---

27 The point in time the knowledge stock in the network has reached its steady state is 61 for WS networks, 55 for ER networks, 45 for BA networks and 32 for networks created with the EV network algorithm.

28 To determine why a node stops trading we define a variable for each node which contains the information whether its unsuccessful trades failed because the respective node had insufficient knowledge or actually its trading partners have insufficient knowledge. The mark colour indicates the average results over a simulation run of 100 time steps.

conclusion that in fact nodes with a high number of links acquire much more knowledge than most of the other agents in the network (Figure 38). They stop trading because their partners have nothing left to offer, however, the difference in diffusion performance between the four network topologies cannot be solely explained by the 'star' argument. In fact, our results indicate a second effect which seems to dominate the diffusion processes in the networks. Very small agents with only few links can only receive very little knowledge and, hence, stop trading first. Considering the sheer number of small nodes in networks with a dispersed degree distribution we have to assume that these small nodes are responsible for disconnecting the network and interrupting knowledge flow. This, in turn, negatively influences the effectiveness of the diffusion of knowledge within the entire network.

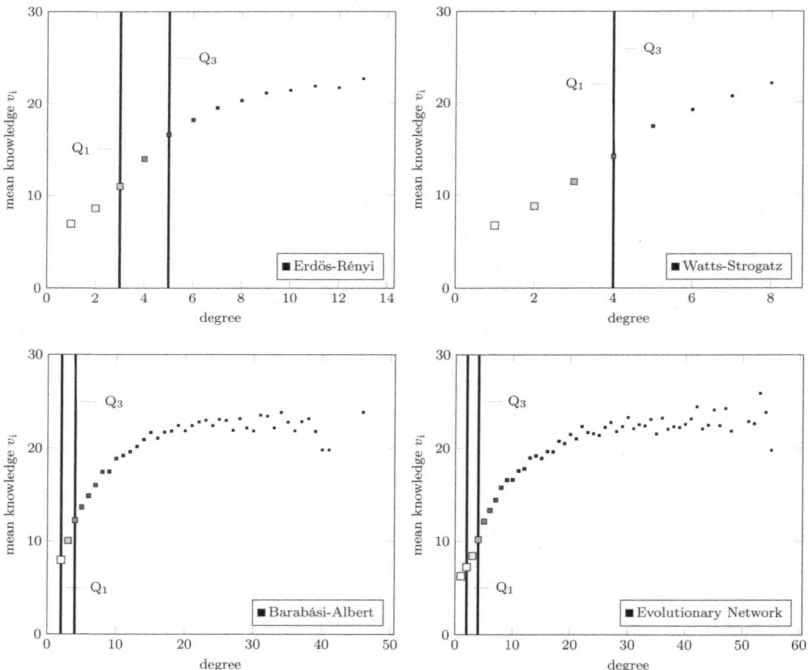

Figure 38: Relationship Between Knowledge Levels and Degree.

To tackle the question whether the absolute degree of nodes in the network is the decisive factor influencing diffusion performance or whether the relative position (i.e. the difference between nodes and their partners) is most important. Figure shows the relationship between the node's knowledge level and its relative positions in the network ω_i. As we see in Figure 38, for the barter trade diffusion process, the relative position of nodes is of decisive importance. For all network topologies, nodes with an advantageous relative position (i.e. with a positive degree difference ω_i between nodes and their partners $\omega_i > 0$ demonstrate a considerably higher knowledge level compared to nodes with fewer links $\omega_i < 0$. This effect is particularly strong for degree differences of $\omega_i = -5$ to $+5$. However, we also see that more extreme degree differences do not increase or decrease a nodes' knowledge levels considerably which again indicates a saturation effect.

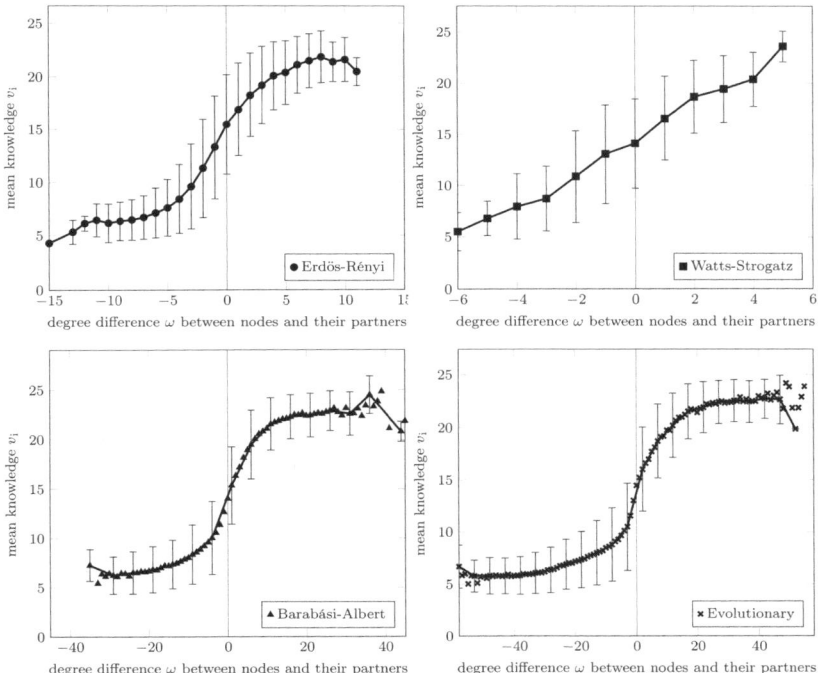

Figure 39: Relative Position and Average Knowledge Levels.

Our results demonstrate that neither path length nor cliquishness are fully sufficient for explaining the performance of knowledge diffusion in networks. Furthermore, other factors, such as the degree distribution, have to be considered in order to fully understand the relevant processes within networks. In contrast to the findings of Cowan and Jonard (2007), our results show that the dissimilarities between nodes - especially for dispersed networks with scale-free structures - can create gaps in knowledge levels. These gaps create a situation where small nodes, which make up the majority of nodes in these networks, do not gain knowledge fast enough to keep pace with the other nodes in the networks. Hence, the many small agents rapidly fall behind and stop trading, which disrupts and disconnects the network and the knowledge flow. Comparing medium-sized nodes and big nodes, however, we also see that stars play a key role in networks.

5.1.3.3   Policy Experiment

From a policy perspective it is important to note that network topologies can be systematically shaped and designed. In other words, the structural configuration can be manipulated in order to increase the knowledge diffusion efficiency of the system. Against the backdrop of section 5.1.3.2 the question arises in which way the system can be manipulated to gain a performance increase on a systemic and on an individual level. Our policy experiment is conducted as a comparative analysis; in which we add new links to an existing network. The general idea behind this is straight forward. We first define three subgroups within the population of nodes and then systematically analyse the effect of adding new links within and between the predefined subgroups.[29]

In this section we present an exemplary policy experiment in which we analyse the effect of six interventions. As a baseline scenario we assume a situation where we have no intervention at all, i.e. the number of links in the network does not change, which in the plots is indicated by the horizontal reference line. This 'no intervention' scenario is used as a reference or control scenario for the actual policy interventions.

- Intervention 1 (stars-to-stars), shows the diffusion performance in a situation in which we equally distributed 20 additional links between stars.
- Intervention 2 (stars-to-medium), shows the diffusion performance in a situation in which we distributed 20 new links between stars and medium nodes.
- Intervention 3 (stars-to-small), shows the diffusion performance in a situation in which we distributed 20 new links between stars and small nodes.

---

29   In the policy intervention, we define *stars* as those 10% of all nodes that have the highest degree centrality, whereas *small* are defined as those 10% of the distribution that have the lowest degree centrality. *Medium* agents are those 80% of the distribution that are neither *stars* nor *small*. To measure the performance of the policy interventions we measure the steady state knowledge stock for every policy after 100 simulation steps and over 500 simulation runs.

- Intervention 4 (medium-to-medium), shows the diffusion performance in a situation in which we equally distributed 20 new links between medium nodes.
- Intervention 5 (small-to-medium), shows the diffusion performance in a situation in which we distributed 20 new links between small and medium nodes.
- Intervention 6 (small-to-small), shows the diffusion performance in a situation in which we equally distributed 20 new links between small nodes.

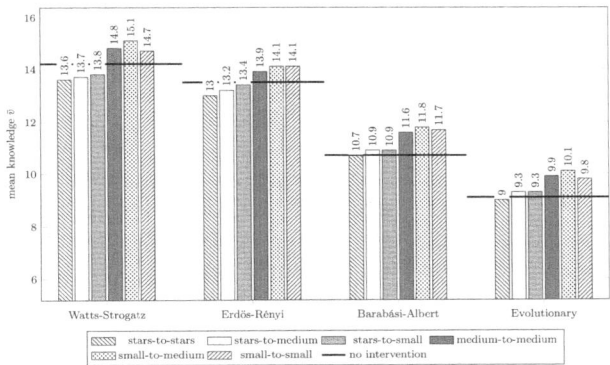

Figure 40: Average Knowledge Levels of Small Agents with Policy Interventions.

As shown in Figure 40, all interventions applied to stars (1, 2 and 3) actually may have a negative (or in the best case marginal positive) impact on network performance. This is interesting because additional links should improve the network's performance as they create new trading possibilities. However, for networks with a symmetric degree distribution these additional links limit the knowledge flow. The explanations for this can be found when we look at the degree distribution of the networks. Although additional links in the network open new trading possibilities they also increase the asymmetry of the degree distribution. This leads to the effects described in subsection 5.1.3.2. On the one hand side small nodes stop trading early because they have to little knowledge. On the other hand, side nodes with a high number of links stop trading because they do not find partners with sufficient knowledge levels. On the overall network level, policy interventions supporting stars or *picking-the-winner* strategies, hence, never seem to be recommendable as they increase the asymmetry of the networks.

If we now analyse all interventions applied to small nodes (3, 5 and 6), it becomes evident that the effect in general is positive (or in the worst case there is no effect). Interventions aiming at small nodes do not increase the asymmetry in the degree distribution, but rather reduce it. This in turn relativize the effects discussed in section 5.1.3.2. Interestingly, the best performing intervention at the system level is not a 'small-to-small' intervention - as one may presume based on the findings of section 5.1.3.2 - but rather interventions creating links between small and medium sized nodes.

While the experiment in Figure 40 analysed the implications of our findings on an aggregated network level, we also have to consider the individual level. Let us start with recalling that the disparate structure of the network itself is only the result of the myopic linking strategies of its actors. In BA and EV networks, one key element of the actors' strategies is preferential attachment, according to which linking up to other high degree actors is likely to increase the individual knowledge stock. Yet this linking strategy leads to the network characteristics discussed above which hinder the diffusion process on the network level, and this, in turn, prevents smaller nodes from gaining knowledge. To put it another way, the limiting network characteristics, which hinder the knowledge diffusion at both the actor and the network level, are actually caused by the behaviour of small nodes which aim to receive knowledge from larger nodes. Hence, in contrast to how small nodes actually behave in BA and EV networks, the optimal strategy for gaining knowledge might actually be to link up with other small nodes.

By looking at Figure 41 we see that this actually holds true. Although on the global scale the linking of small and medium nodes performed best, at the individual level and, more specifically, for small nodes, the best strategy to gain an individual optimum is to connect with nodes that are most similar to them, e.g. to follow a strategy inspired by structural homophily.

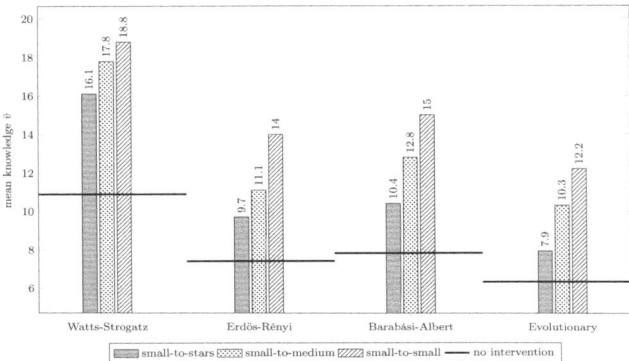

Figure 41: Average Knowledge Levels of Small Agents with Policy Interventions.

5.2   The Importance of Consumer Networks

Building on the previous analysis, the following section adds a network perspective for consumers. For this, we first implement boundedly ration consumers (subsection 5.2.2). In subsection 5.2.3 we then introduce consumer networks.

5.2.1   Introducing Remarks

One of the central assumptions in our baseline model is the assumption of rational consumers, meaning that all consumers always have full information about all firms, their products and the particular characteristic but also on their own preferences for products. The following chapter will stepwise extend the existing baseline model from chapter 4 and analyse the changes in the behaviour of the simulation model. If not stated otherwise, the parameter setting of this chapter's scenarios follows the parameter setting from chapter 4.

We start by limiting the information of consumers about firms and their products. Consumers in this scenario are not automatically informed about all products and their characteristics but are randomly informed by only a small subset of firms.

The assumption of full information in our baseline model may be to some extend justifiable in small markets with a manageable number of firms and products consisting of well-informed consumers actively seeking for new information. Consumer capabilities in large markets are limited, which has strong effects on the innovation process of firms (Valente 2009, 2012). So instead, in this section we follow Herbert Simon's idea of bounded rationality of economic actors

(Simon 1955, 1959, 1972) which is also one of the essential key elements of evolutionary economic theories. This concept has been successfully integrated in models of innovation it is often only applied on the supply and, hence, on the firm side. Consumers, instead, are often reduced to the same hyper rational and fully informed entities evolutionary economics deliberately repelled over the last decades. As for example Valente (Valente 2009, p. 1) notes:

> "This is a curious since the most diffused defence of the perfect rationality hypothesis is the 'as if' used by Milton Friedman, sustaining, in an ironic twist, and extremist version of the selection concept: no matter what people actually do, only the best will survive the competitive test, and therefore economists can focus on the only (supposedly optima) surviving behaviour, whether it stems from design or purely by chance."

Consumers are not subject to competitive pressure and, thus, there is rarely a selection process driving some consumers out of the market which makes the simplifying assumption of rationally and perfectly informed consumers somewhat hollow. Instead, it is clear we have to accept consumer's imperfection and consequently analyse the implication these imperfections have on innovation processes.

Building on this, in a second scenario (section 5.2.3) we add consumer networks to the simulation model. Although networks have long been identified as an interesting field of studies in innovation economics the focus of the analysis has been without doubt on the supply side on networks of firms. Most of the literature on networks in economics is focused on the importance and effects of firm networks and the diffusion of knowledge and technologies (see also chapter 5). The particular role of consumer networks and the way consumers affect each other's decisions and the resulting effects on the innovation process is still underexplored (see for example Babutsidze 2012, Cowan et al. 1997, Valente 2009, Valente 2012 for interesting exceptions).

Consumer networks have without doubt a more informal character as firm networks. The possible connections are based on personal friendship, family relations, etc. They serve as an important source of information, but may determine not only what consumers know but also what they want (see for example Banerjee et al. 2013, Fogli, Veldkamp 2014, Zeppini, Frenken 2015). Following this idea we introduce in subsection 5.2.3 two distinct effects of consumer networks which are designed to emphasize the role of consumer networks for the analysis of innovation processes. First, we analyse the role of consumer networks as a source of information. Building on the scenario from section 5.1.2 we assume in this setting that consumers can acquire information about products of firms through individual links to other consumers who share the information about products they purchased. We analyse hereby how different network topologies affect the outcome of the simulation. In a second scenario consumers are assumed to influence each other's preferences. Instead of just exchanging information about firms and their products in this scenario consumers try to impose their preferences for product peculiarities on other consumers (see for example Swann 1999, Witt 2001a, Ciarli et al. 2008, Babutsidze 2012, Valente 2012).

5.2.2 Bounded Rationality of Consumers

We assume for the next set of experiments that consumers are not automatically informed about the product offered. Instead, firms have to actively send the information about their product via advertising activities. Every time step firms send information about the produced products to a fixed number of randomly chosen consumers defined by the parameter ($\varepsilon \in \mathbb{N}$). However, every successful innovation which changes the product will make old information useless and all consumers forget information about this particular firm's product. Additionally, we assume that consumers have only limited capacities to store information. Consequently, there is a constant probability of 10% to forget every piece of information per time step.

To analyse the key changes introduced by the limitation of knowledge on the consumer side, we first show in Figure 42 (left hand side) the average number of firms and the average CPM level achieved (right hand side) for different heterogeneity settings. We further assume for this experiment that firms randomly choose three consumers per time step to inform them about products ($\varepsilon = 3$).

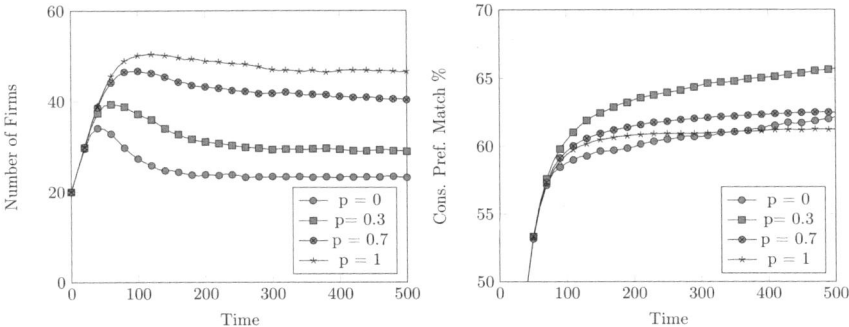

Figure 42: Number of Firms (left) and CPM Levels (right) with Limited Consumer Information.

Figure 42 shows that with an increase in the heterogeneity of demand (increasing heterogeneity parameter p) more firms can survive in the market. However, compared to the result from section 4.3 it becomes evident that the number of firms has increased in case of homogeneous demand and decreased in case of heterogeneous demand. Additionally, especially for markets with a low level of heterogeneity ($p = 0$ and $p = 0.3$) the number of firms first increases and then, after a peak around time step 100, decreases and finally reaches a stable level. These patterns give evidence that two additional but distinct effects are relevant in the scenario of limited consumer information, i.e. on the one hand the effect of additional segmentation and on the other hand the effect of a reduced segmentation of markets.

In the standard scenario the segmentation of markets (both static and dynamic) was solely created by the heterogeneity of demand. In a scenario of limited consumer information, instead, it is of vital importance that consumers have the information about a firm's product. In a situation several products compete it is possible that although one product may be considered best in terms of the offered product characteristic, other firms gain market shares as consumers perceive this product best considering the limited information they have. This creates additional space for firms in the market explaining the positive shift in the number of firms in the situation of homogenous demand.

This additional segmentation is induced by the scarcity of information and takes effect especially in the case of homogeneous demand. As a result, the results show not only an increased number of firms compared to the standard scenario from section 4.3 but also more innovation activities by firms (see Figure 43). While in the baseline scenario the innovation activities of firms are limited to only a few firms engaging in radical innovation, in case of homogenous demand, the limitation of consumer information creates the same pattern as seen for heterogeneous demand in section 4.3. In the beginning of the simulation, firms engage especially in incremental innovation. At the later stages this pattern changes and most firms follow radial innovation strategies.

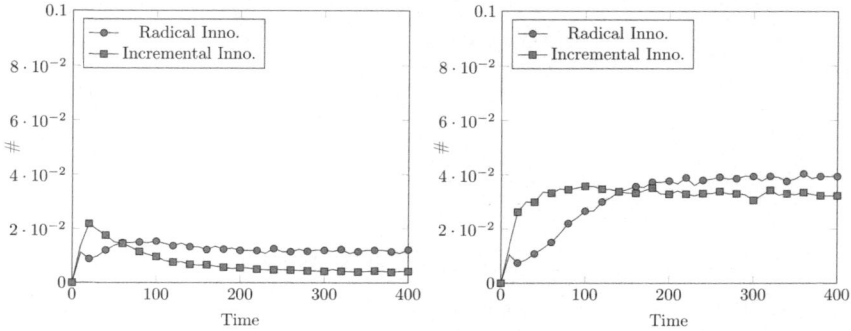

Figure 43: Innovations During a Simulation Run with Limited Information (left: $p = 0$, right $p = 1$).

However, the limitation of consumer information also decreases the chance for successful innovation which explains the reduced number of firms in this scenario compared to a situation of fully informed consumers. In the case of limited consumer information, the chance of a successful radical or incremental innovation is not only determined by the particular product characteristic but also by the ability to inform consumers about a firm's superior product. This reduces the chance for successful innovation and in the long run reduces the segmentation of markets.

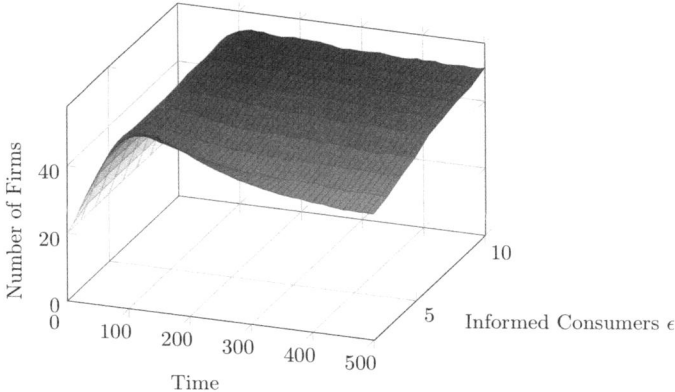

Figure 44: Reduced Segmentation in Heterogeneous Markets.

Finally, for both effects the results show that their impact depends on the potential for technological progress. Especially for the latter, our results indicate that the impact is high when the potential for technological progress is low. At early stages of the simulation, market incumbents are often replaced by competitors with better products. This in turn limits the time the incumbent can spread the information about its product which hinders a strong market position. When the potential for technological progress decreases, however, long-established firms emerge, dominating the market even in heterogeneous markets. At this stage of the simulation the limitation of information actually limits the possibilities of firms to enter the market and steal market shares from competitors. This effect is also depicted in the next figure. Figure 44 shows the number of firms in a market with heterogeneous consumers ($p = 1$) over time and for different values of ε. Even in heterogeneous markets, the limitation of information reduces the segmentation of markets when consumers are badly informed ($\varepsilon < 5$).

The effects described so far also influence the results for the average CPM levels of consumers which is depicted in Figure 42 right hand side. We firstly see a major drop compared to the results of the baseline model (see 4.3.2). While in the baseline model with fully informed consumers the average CPM level ranges between 81% for homogeneous demand and 78% for heterogeneous demand, in case consumers are not fully informed about all products, the CPM levels drop and show also a different ranking. In fact, markets with the highest CPM levels in this scenario are markets with medium levels of consumer heterogeneity.

Finally, in Figure 45 we show the effect of different values of ε on the average CPM level of consumers after 500 time steps of the simulation in a market of heterogeneous consumers ($p = 1$). This scenario has also relevance in terms of possible policy implications. One of the central measures to foster innovation are policies designed to build awareness among consumers e.g. by information campaigns, advertisement etc. (see for example Edler, Georghiou 2007, Edler 2009). Figure 45 shows a clear positive relationship between the level of scarcity of information (indicated by our parameter ε) and the average CPM levels achieved in the simulation. Only a small number of consumers informed ($\varepsilon > 2$) is necessary to establish an effective market, producing products matching the demand. Additionally, the results show that the effect of more consumers informed is always positive but also shows diminishing marginal effects.

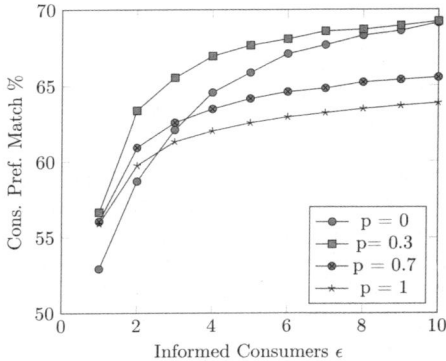

Figure 45: Average CPM and the Number of Informed Consumers.

As a brief summary of the results we can state that the limitation of consumer information is an important aspect which introduces major changes which without doubt have to be included in the analysis of innovation processes. The scarcity of information on the consumer side creates additional heterogeneity amongst consumers and, thus, changes the way firms engage in innovation activities.

In more detail, the simulation results indicate two distinct effects, i.e. on the one hand the effect of additional segmentation and on the other hand the effect of a reduced segmentation of markets. However, the results show that this effect is strongly dependent on the technological opportunities of markets. At early stages of the simulation, the limitation of information lead to additional segmentation of markets and, hence, more innovation. When the potential for technological progress decreases, the limitation of information actually limits the possibilities of firms to enter the market which results in the emergence of long-established firms dominating the market even in heterogeneous markets. Finally, the results indicate that information represents a crucial factor for the development of successful markets.

5.2.3   The Effects of Consumer Networks

In the following section consumer networks are introduced into the model. We hereby aim to study the role of different consumer network topologies with distinct network characteristics on the market processes. In more detail, we apply the four network algorithms already introduced in section 5.1, namely the Watts-Strogatz (WS), Erdös-Rényi (ER), Barabási-Albert (BA) and Evolutionary (EV) networks). Figure 46 shows an example of a consumer network created by the BA algorithm with 100 consumers and 300 links indicating the individual relationships between consumers (red).

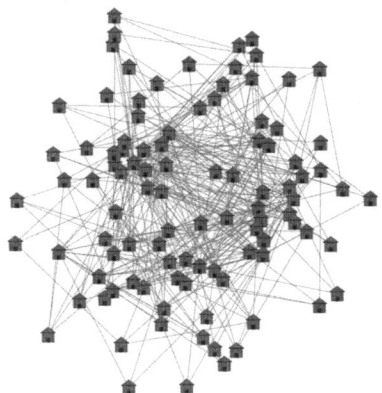

Figure 46: Example of Consumer Networks.

As in the previous analysis, we apply these networks to obtain structurally different consumer networks in the simulation. In particular, these networks differ in terms of path length, cliquishness and the respective distribution of degrees.

Table 4 provides some important information about the network characteristics of our four network topologies.

Table 4:   Network Characteristics of all Four Network Topologies.

	Watts-Strogatz	Erdös-Rényi	Barabási-Albert	Evolutionary
Path length	3.36	2,75	2.56	2.33
Clustering Coefficient	0.38	0.06	0.16	0.41
Degree Distribution	not dispersed	not dispersed	dispersed	dispersed

While WS networks are characterised by a high path length and a high clustering coefficient ER network show considerable lower levels of path length and a lower clustering coefficient. Both networks show a relatively small variance in the degree distribution. In contrast to this, BA and EV networks exhibit a dispersed degree distribution following a power-law approximately. Networks with the lowest path length and the highest clustering coefficient are EV networks.

Our first experiment builds on the previous scenario from section 5.1.2 and is designed to analyse the effect of different consumer network topologies on the emergence of markets. In more detail, in this scenario consumer networks act as a source of information about products in case of limited consumer information. Accordingly, we assume for this scenario that with a constant probability of 30% each time step every consumer sends out the information about the products it has purchased in the last period to direct link neighbours.

Figure 47 shows the number of firms in a situation of heterogeneous demand for the four different types of consumer networks as well as for a situation without consumer networks. As we can see, the differences between the four consumer networks are only marginal. All four network topologies seem to produce very similar results. However, compared to a situation without a consumer network the number of firms increases.

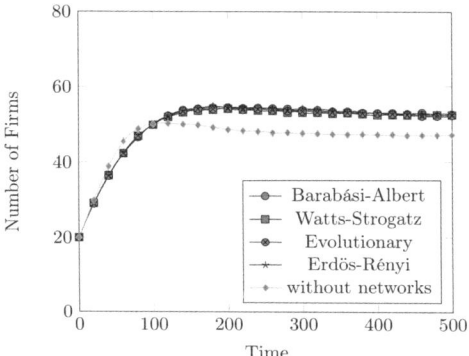

Figure 47: Number of Firms with by Consumer Networks.

The same pattern can be observed if we look at the average CPM levels as depicted in Figure 48. The average CPM levels for all four consumer network topologies stay identical over the analysed time. However, in line with the increased number of firms (see Figure 47) consumer networks as the source of information about possible sellers increase the average CPM considerably. This leads to the question why in contrast to the findings for example by Fogli and Veldkamp (2014) and Zeppini and Frenken (2015) different network characteristics in our simulation model do not lead to changes in the results.

One possible explanation is that consumer networks in our simulation only give additionally information about sellers adding to the existing information obtained by advertising activities by firms. However, this factor alone cannot fully explain the results because our analysis shows that even if firm information advertisement is very low (e.g. informing only one randomly chosen consumer per time step, $\varepsilon = 1$) the results hold.

A second explanation lies in the factor of consumer heterogeneity. In case consumers exhibit heterogeneous preferences for products the information about possible sellers by a consumer's friend is simply of less importance. The information consumers receive is based on the purchasing decision of consumers and, hence, is based on their preferences. As these preferences differ in the network, the information about a friend's products may not suit my own preferences. However, while this may be true for heterogeneous demand, for markets with homogeneous demand this explanation fails as all consumer share the same preferences which makes information by other consumers about products of great value.

Our results indicate that also in markets with homogeneous demand the differences between the four network topologies are only marginal. In this case, only a small number of firms exists and, hence, there is only a small number of firms innovating. As a consequence of the relatively dense network, all consumers are quickly informed about the best possible product independently from the network topology.

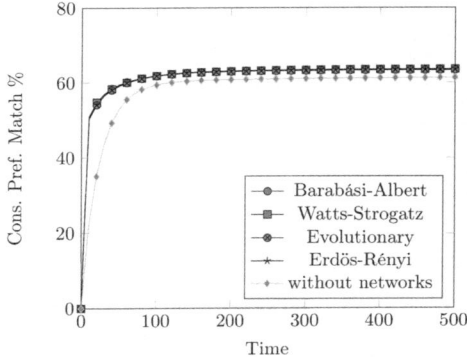

Figure 48: Consumer CPM Levels with Consumer Networks.

Inspired by these results we added in Figure 50 two additional networks which are based on the concepts of *homophily* and *heterophily* (see for example Rogers 2010). Instead of just randomly connecting consumers, these two algorithms assume that consumers with the most *similar* respectively *different* preferences connect to form a network.[30] Table 5 and Figure 49 provide some details about the resulting network characteristics.

Table 5:   Network Characteristics of Homophily and Heterophily Networks.

	Watts-Strogatz	Erdös-Rényi	Barabási-Albert	Evolutionary	Homo-phily	Hetero-phily
Path length	3.36	2,75	2.56	2.33	2.752	2.712
Clustering Coefficient	0.38	0.06	0.16	0.41	0.087	0.028

---

30   In more detail: we compute for both algorithms the distance between all pairs of consumers determining the most similar and the most different possible partners. In a second step, each consumer is asked to connect to the most similar or the most different partner until the network reaches 300 links.

Notably, both additional networks show medium path length and a degree distribution similar to random networks. Interestingly, the clustering coefficient for homophily networks is considerable higher than for heterophily networks.

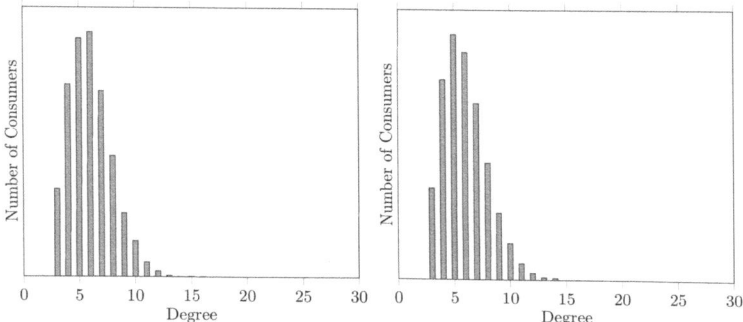

Figure 49: Degree Distribution of Homophily (left) and Heterophily (right) Consumer Networks.

As Figure 50 shows, looking in more detail at the results for the six different networks, although the *homophily* and the *heterophily* network do not exhibit outstanding network characteristics, they still are networks with the best, respectively worst, average CPM levels. In other words, for networks between heterogeneous consumers, it appears that the overall network topology is of less importance and the more qualitative information which consumers are connected is the decisive factor for the network's performance.

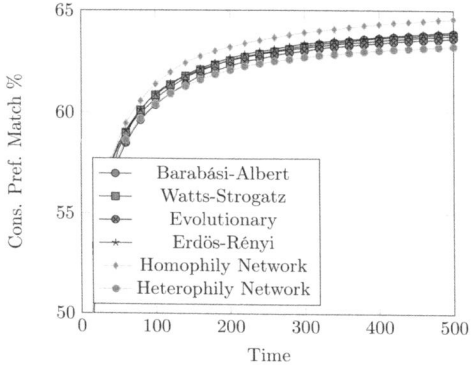

Figure 50: CPM Levels for Homophily and Heterophily Consumer Networks.

Although the network topology on average did only result in marginal changes in the CPM levels we are nevertheless interested in the particular effects on the individual consumer level of our four main network topologies. Inspired by the results from section 5.1 we show in Figure 51 the relationship between the number of friends a consumer has and the average CPM level of consumers with a particular number of friends for our four standard network topologies (i.e. BA, EV, WS and ER networks).

Although the overall network topology seems to be of less importance for the average CPM levels of consumers, the particular number of friends does indeed affect the match between demand and supply on an agent level showing a positive relationship. Especially in BA and EV networks consumers with only a few number of links show a significant lower match as consumers with a high number of links which is in line with the results from chapter 5. However, because consumers with a high number of links simultaneously show a relatively high CPM, on average the difference between the networks topologies is only marginal.

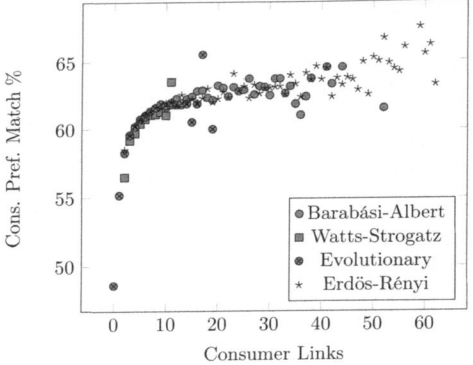

Figure 51: Degree and CPM Levels for Different Network Topologies.

This is also evidenced by Figure 52 where we depict the minimum CPM levels achieved in the simulation for different network topologies over time. Here the difference between different network structures becomes clear. While on average consumers in the different networks perform equally, consumers in networks with a high asymmetric degree distribution i.e. in EV networks are of high risk to perform poorly and consumers in networks with a high symmetrical degree distribution (i.e. WS networks) achieve considerable higher CPM levels.

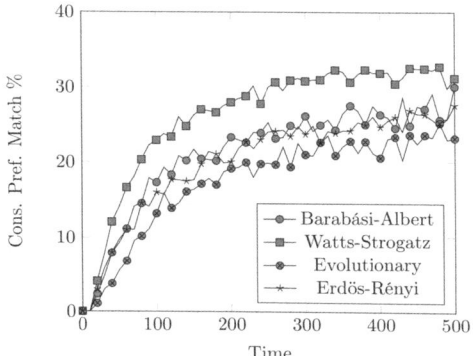

Figure 52: Minimum CPM of Consumers in Case of Homogeneous Demand.

As a second part of our analysis in this section, we analyse in the following experiment a further effect of consumer networks. Consumer networks do not solely act as sources for information about products and their characteristics. Additionally, consumer networks also have direct influence on the preferences of consumers (see for example Swann 1999, Witt 2001a, Ciarli et al. 2008, Babutsidze 2012, Valente 2012). To implement the mutual influences amongst consumers embedded in networks we assume for the following that consumers copy the preferences from a randomly chosen neighbour with a constant probability of 30%.

Figure 53 depicts the resulting relative demand distance between consumers for an initially fully heterogeneous demand. All four network topologies show a major decline in the relative distance over time, indicating a system-wide convergence of demand. However, we also see that there are fundamental differences which depend on the particular network algorithm used.

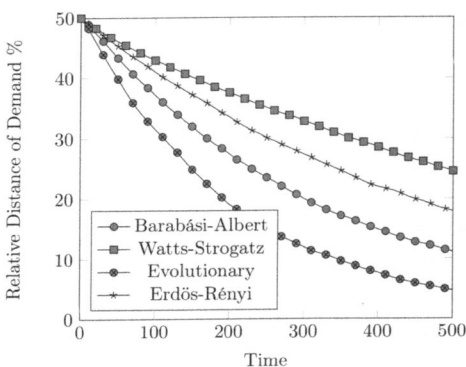

Figure 53: Relative Distance Between Consumers over Time.

In networks with a small average path length and a dispersed degree distribution (BA and EV networks) the relative distance between consumers decreases considerable faster than for networks with a high path length and more uniform degree distribution (WS and ER networks). One possible explanation for this is an effect also relevant for the so called *majority illusion* as for example studied by Lerman, Yan and Wu (2015). In networks with a dispersed degree distribution preferences of stars (i.e. consumers with a high number of links) are systematically overrepresented in their local neighbourhood. The preferences of these stars quickly diffuse through the network causing the relative distance between consumers to converge quickly.

This also explains the patterns in Figure 54 where we show the number of firms and the average CPM levels achieved for different consumer networks. With a decreasing relative consumer distance, markets approach a situation of homogeneous demand. This, in turn, reduces the number of firms in the market (Figure 54 left hand side) and simplifies the search problem of firms which also leads to higher CPM levels of consumers (Figure 54 right hand side).

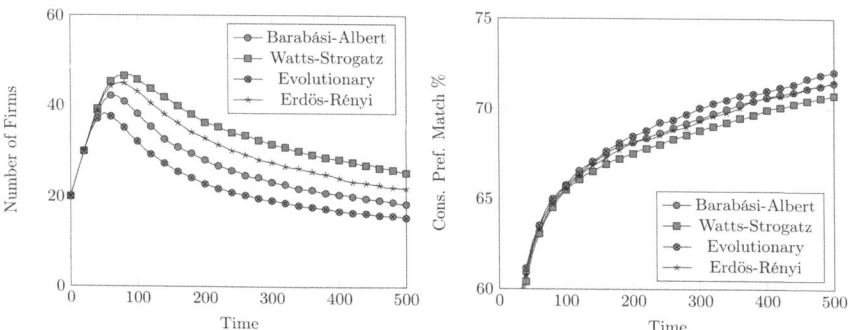

Figure 54: Number of Firms and CPM Levels Over Time.

Notably, this pattern is in contrast to our observation from chapter 4 where the model produced right-tailed log-normal distributions for $p = 0.7$ and standard normal distributions for $p = 1$. However, as stated before, markets show right-tailed log-normal distributions at the time of birth of an industry and becomes becomes more symmetric over time (Cabral, Mata 2003). This implicates that heterogeneity on markets actually is increasing and not converging as indicated in this experiment.

Summing up this section's results, we see that introducing consumer interactions via networks without doubt stresses an important issue in the debate on the role of consumers in the innovation process. Our experiments show that consumer networks are an important source of information but also a desicive factor for the convergence of preferences. Especially the individual position of consumers, i.e. the number of connections and the individual characteristics of connected consumers determines whether a consumer finds the right products and, hence, can be a good target for policy measures.

5.3 Discussion

This chapter introduced a network perspective to the analysis. We divided the chapter into two separated building blocks. On the one hand, we analysed how different network topologies affect the network diffusion within firm networks. On the other hand, this chapter added consumer networks in which consumer share information and influence each other's purchasing decision.

Economic actors in today's economies become without doubt more and more connected and interlinked. Most scholars in the field of interdisciplinary innovation research would agree that "networks contribute significantly to the innovative capabilities of firms by exposing them to novel sources of ideas, enabling fast access to resources, and enhancing the transfer of knowledge" (Powell, Grodal 2005, p. 79). However, we still face the question which network topologies are most effective and efficient in enabling the diffusion of knowledge. In this context, it has been frequently argued that small-world networks – characterised by short path lengths and high cliquishness – demonstrate superior knowledge diffusion properties.

Our analysis showed that, in line with the findings of Cowan and Jonard (2007), a highly asymmetric degree distribution actually has a negative impact on the overall network performance. However, different to the explanation given by Cowan and Jonard, we found that this negative effect cannot be explained (solely) by the existence of stars that rapidly acquire knowledge and so interrupt the trading process. Our results show that neither do stars acquire more knowledge than most of the other agents, nor do they stop trading earlier. Our findings indicate that stars trade longer than 70% of the nodes and only stop trading after most of the knowledge already has diffused throughout the network. A group of agents that actually has a very low level of knowledge and stops trading long before most of the knowledge already diffused throughout the network is the group of very small, inadequately embedded agents. Notably, our results support the idea that it's actually the dissimilarity in degree distribution itself. This effect also holds for dense networks and, hence, for networks in which small nodes still have a high number of links.

Finally, we conducted a policy experiment aimed to show the potential consequences of our findings. The results indicate that in all four networks the group which benefited most from an increase in its links is the group of very small agents. Our results clearly show that in networks with a skewed degree distribution, not the stars, hinder knowledge diffusion but very small agents do. Summing up, our analyses lead us to the conclusion that first, a highly skewed degree distribution negatively influences diffusion of knowledge that is exchanged in a barter trade process. Second, very small agents are the bottleneck for the efficient diffusion of knowledge throughout the networks.

As stated before, although introducing heterogeneous consumer preferences in chapter 4 already presented a huge step towards a more precise representation of consumers it still simplifies the behaviour of consumers in many aspects. If we want to understand the causal relationships between demand and innovation, we need to consider consumers as heterogeneous but also bounded rational beings embedded in networks.

We started by implementing the concept of limitedly informed consumers which lead to unexpected results. Changing the way consumers are informed creates first of all information scarcity on the consumer side. This scarcity of information creates additional heterogeneity amongst consumers and, with this, changes the way firms engage in innovation activities. In more detail, the scarcity of information creates two additional effects which fosters respectively hinders the segmentation of markets and with that fosters respectively hinders innovation.

This effect is strongly dependent on the technological opportunities of markets. At early stages of the simulation, the limitation of information lead to additional segmentation and, with this, more innovation. As the potential for technological progress decreases, the limitation of information actually limits the possibilities of firms to enter the market which results in the emergence of long-established firms dominating the market even in heterogeneous markets. This also has strong effects on the performance of markets in terms of the achieved CPM levels, i.e. the match between what consumers want and what firms' products provide. In fact, neither fully homogeneous markets nor fully heterogeneous markets are able to produce products on the same level as markets with full information. In contrast, markets with a medium level of consumer heterogeneity performed best.

Subsection 5.2.3 introduced consumer networks and analysed two distinct effects within networks. Building on the scenario from section 5.1 we see that networks are an important source of information not only for firms but also for consumers. However, our results also reveal that the value of information in networks of heterogeneous consumers depends on the individual characteristics of the consumers. In networks of heterogeneous consumers, information from consumers with different preferences has less value than information by consumers with similar preferences. As a consequence, network characteristics such as path length, cliquishness or the distribution of links has only a minor effect on the outcome of the simulation. Instead, aspects of *homophily* and *heterophily* as already mentioned by Rogers (2010) are of greater influence on an average scale. This result also puts into perspective the outcomes of chapter 5 in which it is assumed that knowledge is something impersonal and equally valuable for all firms.

Looking in more detail at the results it becomes evident that the average CPM levels are not sufficient to analyse the performance of different network structures. In fact, on an individual level, the number of links strongly affects the match between what consumers demand and the products consumers buy. Especially in BA and EV networks consumers with only a few number of links show a significant lower match as consumers with a high number of links. This fact is also relevant from a potential policy side. Policy measures designed to foster the creation and diffusion of innovation, thus, should especially aim at less connected consumers.

In a second experiment, subsection 5.2.3 introduced that consumers may actually influence each other's preferences. Instead of just exchanging information about firms and their products, in this scenario, consumers try to impose their preferences for product peculiarities on other consumers (see for example Swann 1999, Witt 2001a, Ciarli et al. 2008, Babutsidze 2012, Valente 2012). Our results show that especially in networks with a small average path length and a dispersed degree distribution (e.g. EV and BA networks) the relative distance between consumers, i.e. the heterogeneity of demand, decreases considerably faster than for networks with a high path length and more uniform degree distribution (e.g. WS and ER networks). Inspired by the work about the *majority illusion* (Lerman et al. 2015) we conclude that in networks with a dispersed degree distribution preferences of stars (i.e. consumers with a high number of links) are systematically overrepresented in their local neighbourhood. This leads to the fast spread of these dominant preferences causing the relative distance between consumers to converge quickly. The quick convergence of markets, however, reduces the variety of firms and products which in turn limits the markets possibilities to adapt to new developments (Saviotti 2001).

6 Bounded Morality of Consumers

The following chapter adds an important aspect of products which is too often overlooked. With every purchase, consumers have to decide whether they want to include possible negative characteristics of products with no direct or only indirect influence on the consumer into their consideration or not. The aim of this chapter is to apply the simulation model of the previous chapters for the analysis of *responsible innovation*. In more detail, we extend the model by adding a second product characteristic which represents possible negative aspects of innovations. Consumers in this scenario also are heterogeneous in the way they consider and evaluate the negative aspects. With this we aim to analyse how *responsible innovations* are created and how they diffuse through the system.

Building on the baseline model from chapter 4 and the model extensions in chapter 5 we divide the analysis of *responsible innovation* into three scenarios. After a brief introduction in section 6.1, subsections 6.2.1, 6.2.2 and 6.2.3 analyse the creation and diffusion of responsible innovations. The extensions and the corresponding results of the simulation are discussed in section 6.3.

6.1 Introducing Remarks

In our final set of experiments, we apply the simulation model for the analysis of the diffusion of responsible innovation. The notion of *responsible innovation* (RI) or *responsible research and innovation* (RRI) is rather new. It can be traced back to the work of different authors and is strongly related to older concepts such as *technology assessment* (see for example Hellström 2003, Owen et al. 2009, Owen, Goldberg 2010, Von Schomberg 2007, Armstrong et al. 2012). Following von Schomberg (2013, p. 19) we can define RRI as:

> "a transparent, interactive process by which societal actors and innovators become mutually responsive to each other with a view to the (ethical) acceptability, sustainability and societal desirability of the innovation process and its marketable products in order to allow a proper embedding of scientific and technological advances in our society."

At its very heart RRI extends the common practise of reducing innovation to something always positive and desirable (Blok, Lemmens 2015). Building also on chapter 4, it becomes evident that innovations have more facets and act within a multidimensional space. Correspondingly, we need to consider that innovations also have a set of negative aspects. In more detail, with Stilgoe et al. (2013) we find four basic dimensions which have to be considered for the innovation process: *Anticipation, Reflexivity, Inclusion* and *Responsiveness*.

In a similar vein as in the previous chapters of this dissertation, we see the consumer side as the driving force for the successful establishment of new and *responsible* products. As a consequence, we aim to analyse the creation and diffusion of responsible innovation triggered by consumers demanding these innovations. The special feature of this scenario is that we extend the model and add a second characteristic of products which can be considered a 'harmful' characteristics (what Block and Lemmens (2015) have called the Faustian aspect of innovation) such as energy or resource consumption, environmental issues, etc.

In this scenario consumers are also heterogeneous in the way they recognize and evaluate negative characteristics. While some consumers put negative characteristics into their consideration, others neglect these characteristics and focus on their individual preference for the main product characteristics as in the previous models. In this way, we analyse the effect of more responsible consumers (i.e., consumers who consider also negative characteristics) on the innovation process. In more detail, we are interested whether there is an optimal market structure which fosters the production of responsible innovation and aim to find possible implications to shift markets towards more responsible innovation.

To implement negative characteristics, we assume for the following that products are characterised by two separated bit strings of identical length, whereas the first bit string determines product characteristics as in chapter 4 and the second bit string represents negative product characteristics. With this it becomes possible, instead of using just one aspect, e.g. the energy consumption or the amount of waste produced for manufacturing, to consider negative characteristics through a multi-dimensional space. While for the first characteristic we continue to assume an individual demand of each consumer (indicated by the heterogeneity parameter p) the potential demand for the second characteristic is fixed for all consumers (indicated by an identical bit string) representing a theoretically socially optimal design of a product.

To mirror the individual acknowledgement of the negative characteristic some consumers evaluate products based on both characteristics (responsible consumers) while other simply neglect the negative characteristic. Responsible consumers weight the individual importance of product characteristics based on the parameter γ, so e.g. if $\gamma = 0$ they also neglect the negative characteristic and for $\gamma = 1$ they fully focus on the negative product characteristic and do not consider any other. Finally, the ratio between consumers not considering negative product characteristics and responsible consumers is given by δ indicating that for $\delta = 1$ all consumers are responsible consumers and for $\delta = 0$ no responsible consumer exist.

As the second product characteristic implemented in the simulation model represents only an abstract simplification of all harmful characteristics of a product we cannot measure the direct negative effects of innovation such as e.g. the increased electricity needed or the waste produced. Instead, we measure the responsibility of a product based on the inverted hamming distance between the socially optimal product characteristic (predefined for each product randomly during the initialisation of the simulation) and the second product characteristic of firms' products (see also chapter 4).

6.2 Model Analysis

The subsequent analysis of simulation results follows the structure of the previous model analysis and starts with the specifications from our baseline model from chapter 4. In the later subsections of this chapter this baseline model will be adapted stepwise to the extensions made in chapter 5. We aim to show the change in the innovative adaption process given consumers with different degrees of our two parameters γ and δ. In particular, we are interested whether there is an optimal combination of both parameters in terms of reduction of negative effects of innovations depending on the degree of the heterogeneity of demand or other factors such as the limitation of consumers' information or consumer networks.

6.2.1 Baseline Scenario

To get a first understanding of the dynamics in the markets Figure 55 shows the match with negative characteristics (left hand side) of products produced and the average CPM levels for degrees of consumer heterogeneity (right hand side) over time. We assume for this experiment that only half of consumers consider the negative product characteristics and evaluate them equally with the other product characteristic, i.e. $\gamma = 0.5$ and $\delta = 0.5$.

As shown in Figure 55 (l. h. s.), we see that over time firms adapt to consumers' preferences and produce products which consider also the negative characteristics (this could for example mean that firms start to look at the energy consumption, the waste problem, employs security and wage etc.). However, we also see major differences for different levels of consumer heterogeneity, indicating a positive relation between the degree of heterogeneity and the resulting match with negative characteristics.

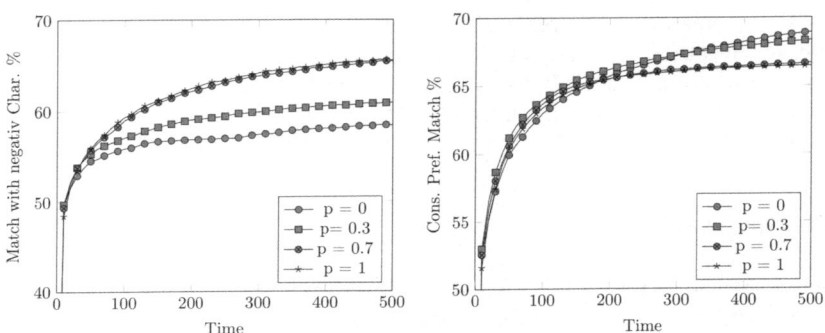

Figure 55: Match with Negative Characteristics and CPM Levels for Different Degrees of Consumer Heterogeneity.

As described in chapter 4, in markets with homogeneous demand there is considerable less innovation than in markets with heterogeneous demand. In this scenario the market can be divided in consumers simply neglecting negative characteristics and consumers equally considering both sets of characteristics. Measuring the average match of products with the theoretical socially optimum, we see that markets fail to produce these products in case consumer preferences for the first product characteristic are homogeneous. As already shown in the previous chapters of this dissertation, in case of homogeneous demand the market does not provide space for segmentation. As a result, only two separated niches, i.e. separated submarkets, one for responsible consumers and one for less responsible consumers emerge. This in turn leads to markets with only few innovations and consequently the average match of products with socially optimal products is only low.

If we increase the level of heterogeneity the clear boundaries between the two niches soften and firms produce a great variety of products. With this, they also produce better, i.e. more responsible products, although the number of consumers considering negative characteristics and the way they consider them stays constant. Additionally, looking at the average CPM levels of consumers, i.e. the match between the characteristics demanded by all consumers (considering negative characteristics or not) and the characteristics of products they purchase there is a major drop for heterogeneous markets which can be explained by great variety of different products demanded. In case of homogeneous demand, innovation efforts of firms are focused. Heterogeneous demand, in contrast, creates a situation in which firms have to focus on small niches and, thus, can only provide compromised solutions which fail to match consumers' preferences.

To analyse in more detail the effects of different degrees of γ and δ we show in Figure 56 the average match of products with negative characteristics achieved in a market of heterogeneous consumers ($p = 1$) with fully informed consumers after 500 simulation steps.

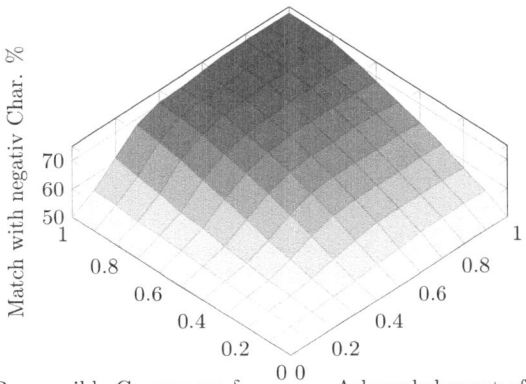

Figure 56: Match with Negative Characteristics in a Fully Informed Market.

In a market with full consumer information both factors γ and δ are of decisive importance. The best match with the social optimum is achieved if all consumers evaluate products only based on the negative characteristic. While this result may seem straightforward, we also see that a situation in which only few responsible consumers exist ($\delta < 0.4$) markets fail to produce products matching the negative characteristics even if these consumers base their decision only on negative characteristics of products ($\gamma > 0.8$). In other words, for responsible innovation to occur, markets need an at least partially balanced composition of both factors γ and δ.

Interestingly, the introduction of negative characteristics has only minor effect on the average CPM levels as depicted in Figure 57. However, we do also see that the maximum CPM levels are achieved in the extreme situation in which all consumers fully consider negative characteristics.

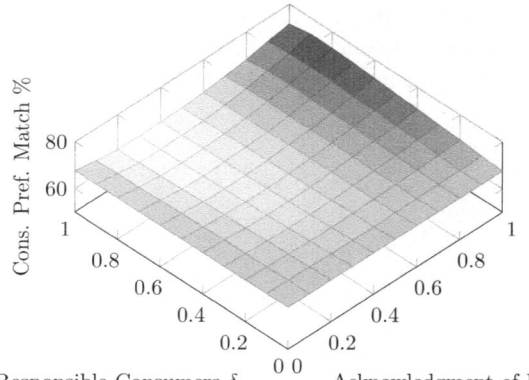

Figure 57: CPM Levels in a Fully Informed Market.

Finally, we are interested in how responsible products are spread over the population of consumers. Figure 58 and Figure 59 show the spread of individual match with negative characteristics between responsible and normal consumers. The match with negative characteristics of products of responsible consumers is mainly defined by their level of acknowledgment of negative characteristics (factor γ). This means that the market produces products matching the demand of responsible consumers even if only a small number of consumers has demand for these products.

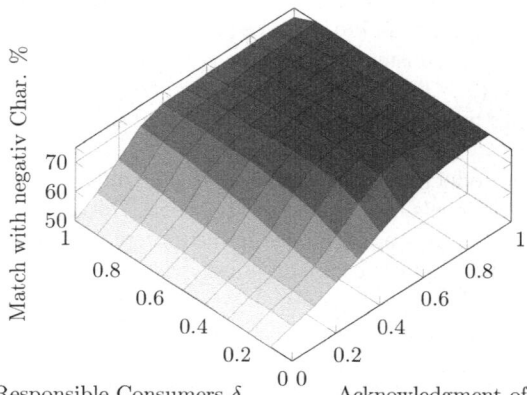

Figure 58: Match with Negative Characteristics of Responsible Consumers.

In the market also the average match of products of normal consumers with the negative characteristics are affected as seen in Figure 59. Although these consumers do not consider negative characteristics, the market produces spill-over effects for a sufficient number of responsible consumers. These spill-over effects are especially strong if normal consumers are outnumbered by responsible consumers (indicated by an increasing factor δ). However, we also see that the spill-over effects in most cases are only marginal, indicating that the market in this situation specialises creating two distinct segments.

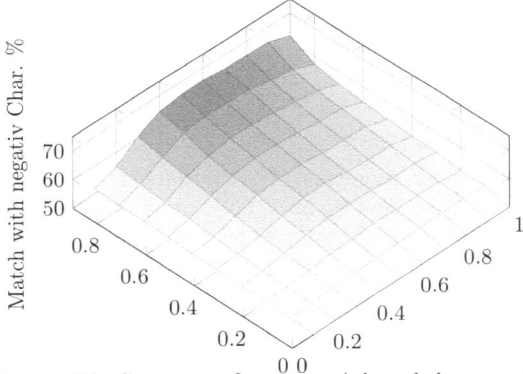

Figure 59: Match with Negative Characteristics of Normal Consumers.

Finally, based on the analysis in chapter 4.3.3 we analyse in the following experiment how consumers may trigger the emergence of new industries. In this scenario consumers are sensitive about the offers and buy products only if they match at least to some extend with the preferences (sensitivity threshold s defines the minimum level of match i.e. the CPM level necessary to trigger sales). Figure 61 shows the results for different ratios of responsible consumers considering only the negative product characteristics (for $\gamma = 1$).

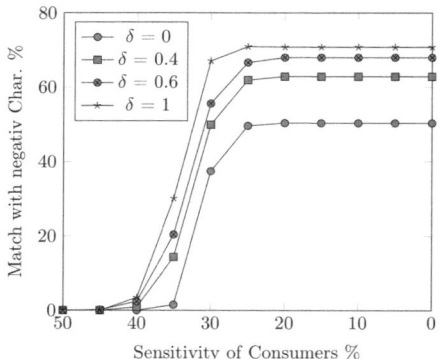

Figure 60: Sensitivity of Fully Rational Consumers.

Independently from the number of responsible consumers the market fails to produce any products for $s > 45\%$. In the small range for $45\% > s > 25\%$ markets start to produce responsible products. In this situation, the number of responsible consumers becomes a significant factor to trigger responsible innovations.

Summing up, it becomes clear that whether the market produces products which match with the socially optimal is not only defined by the number of consumers considering these characteristics but also on the degree of consumers' heterogeneity. In homogeneous markets, two distinct segments emerge, one for responsible consumers and one for consumers who focus on their individual preferences. Only in heterogeneous markets, the dynamic segmentation creates enough innovation to push forward the development of responsible products.

However, as also stated before, the assumption of fully rational and fully informed consumers is a limitation of the simulation model and represents only an extreme case which applies for small markets and consumers actively searching for products. Consequently, we analyse in the following subsections how the aspect of boundedly rational consumers may change the results gained so far.

6.2.2 Responsible Innovation and Limited Information

In this sections' experiments, consumers have only limited information about firms' products and their characteristics. Similar to the analysis in chapter 5.2.2 we assume that consumers are not automatically informed about products and firms send information to three randomly chosen consumers per time step.

As Figure 61 shows, in this scenario, the average match of products with negative characteristics has its maximum if 60% of the consumers are responsible ($\delta = 0.6$) and evaluate both characteristics equally ($\gamma = 0.5$) which is in contrast to the results gained in the previous subsection. To understand this phenomena, we have to recall that in case of limitedly informed consumers, homogeneous markets fail to produce matching products (see chapter 5). For increasing values of γ and δ the market approaches a situation where all consumers only consider the negative characteristic and, consequently, for higher values of γ or δ the market approaches a situation of homogeneous demand. This leads to less firms and, hence, less innovation in the market.

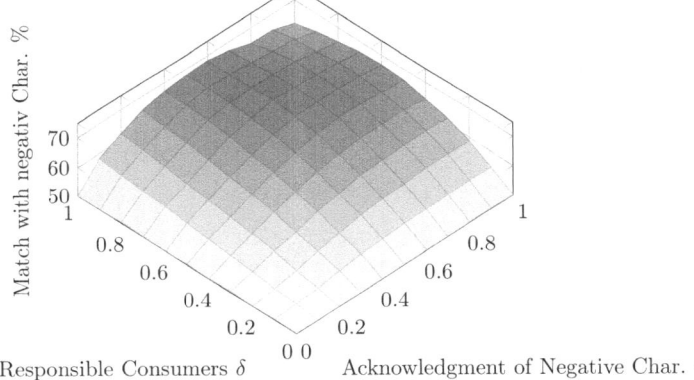

Figure 61: Match with Negative Characteristics with Limited Information.

This explanation is also evidenced by the results depicted in Figure 62. For increasing values of γ or δ the number of firms considerable drops, indicating that in this situation the market produces established firms as market leaders. This market structure leads to a situation where only few firms innovate which reduces the creation and diffusion of responsible products.

As a result, we see a responsibility gap which appears if consumers focus only on negative product characteristics. To put this another way, our results in this scenario indicate an optimal composition of markets. Clearly, if nobody wants responsible products firms will not produce them. At the same time, if everybody only considers negative characteristics and neglects his individual preferences, the result is a homogeneous market. Homogeneous markets, especially in case of limited consumer information, however, reduces the market's possibility to be innovative. Instead, temporally monopolies emerge which limits the innovative dynamics seen in heterogeneous markets.

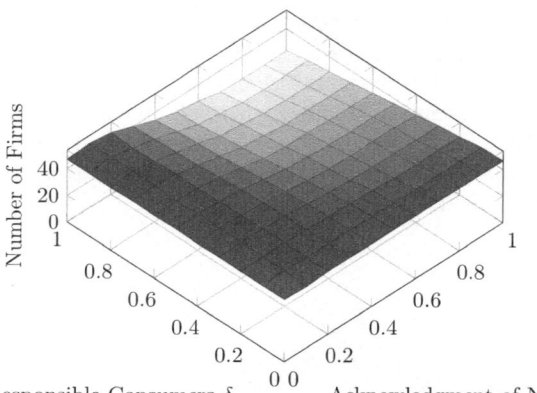

Figure 62: Number of Firms in Case of Limited Information.

Finally, our results support the idea that a scarcity of information also generates more spill-overs. As depicted in Figure 63, in case of information scarcity products purchased by normal consumers (i.e. consumer not considering negative characteristics) match better with negative characteristics indicating an overall positive shift compared to the results in a situation of fully informed consumers. This effect can again be explained by the reduced segmentation of markets. Less segmentation leads to a smaller number of firms and, hence, more normal consumers buy their products from firms occupying niches with responsible consumers.

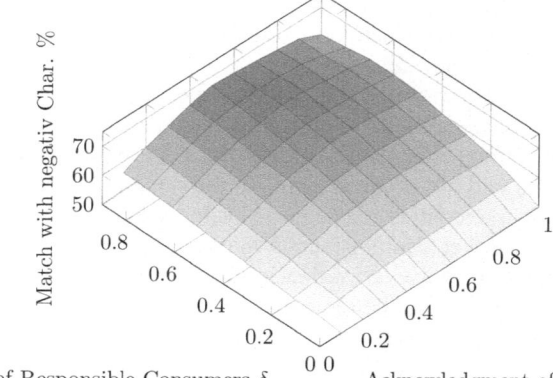

Figure 63: Match with Negative Characteristics of Normal Consumers with Limited Information.

The scarcity of information also affects the results for different levels of consumer sensitivity (see Figure 64). In contrast to the results from the beginning of this section, especially for sensitive consumers ($s > 30\%$) markets with only responsible consumers results in a high match with the theoretical social optimum. Only for low levels of s the responsibility gap pictured in Figure 61 becomes relevant. Additionally, the figure shows that the difference between markets without responsible consumers $\delta = 0$ and markets with only responsible consumers $\delta = 0$ adds up to more than 50% for consumers with a sensitivity threshold $s = 30\%$.

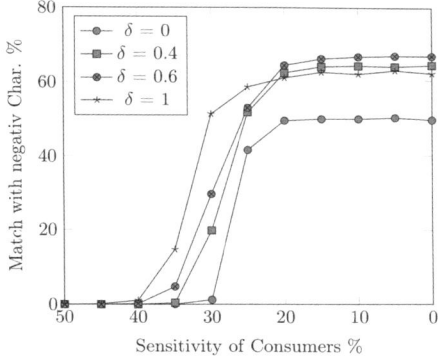

Figure 64: Sensitivity of Boundedly Rational Consumers.

The experiments presented in this subsection illustrate how the dynamics in the markets changes if we abandon the straight and limiting assumption of perfectly rational and informed consumers. While one might expect that the best products are produced if all consumers only consider the negative product characteristics, the market dynamics underlying the creation and production of new products interestingly create a situation where less responsible products are produced than in a situation where some consumers consider negative characteristics and some do not.

However, as stated before, we also have to account for consumer networks which act as an important source of information, to get a full picture of the relevant processes

6.2.3 Responsible Innovation and Networks

For the following experiment we assume similar to the experiments in chapter 5.2 that consumers are embedded in a consumer network of 100 consumers and 300 links. To create the networks for the first experiment the Barabási-Albert network algorithm is applied. In Figure 65 we show that the average match with the social optimum for all combinations of γ and δ increases compared to a situation without consumer networks as in subsection 6.2.2. However, we also see that in this scenario the responsibility gap vanishes and the most responsible products are created for high values of γ and δ, i.e. in a situation where most consumers fully consider negative product characteristics.

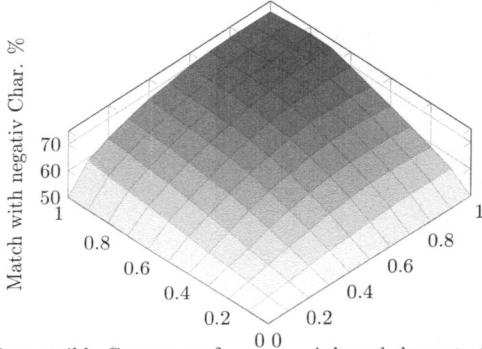

Figure 65: Match with Negative Characteristics with Networks.

Additionally, networks have major effects on the level of spill-over effects in the simulation. While in the previous two settings of this section products of normal consumers showed only a low match with negative characteristics, consumer networks show to raise considerably the match of normal consumers' products with negative characteristics.

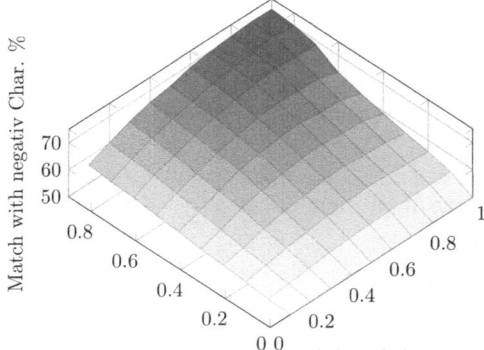

Figure 66: Match with Negative Characteristics of Normal Consumers with Networks.

Finally, to get an in-depth understanding of the importance of network topologies for the creation and diffusion of responsible products we examine a scenario in which we applied different network topologies (see also chapter 5.2). In this scenario consumers are assumed to influence each other's preferences, trying to impose their preferences for product peculiarities on other consumers (see for example Swann 1999, Witt 2001a, Ciarli et al. 2008, Babutsidze 2012, Valente 2012).

In Figure 67 we show the development of products' matches for different networks topologies. Although the differences over time are only minimal, we still can identify major differences indicating a causal relationship between the degree distribution of the particular network topology and the achieved match with negative characteristics of products. Networks with a dispersed degree distribution, e.g. EB or BA networks perform considerably worse than networks with an evenly spread degree distribution, e.g. ER or WS and the two consumer networks (see also chapter 5 for a detailed description on each network topology). Interestingly, both consumer networks (the homophily and heterophily network) show the same results, indicating that in contrast to the scenario in chapter 5.2.3 the preference difference between connected consumers does not affect the simulation results.

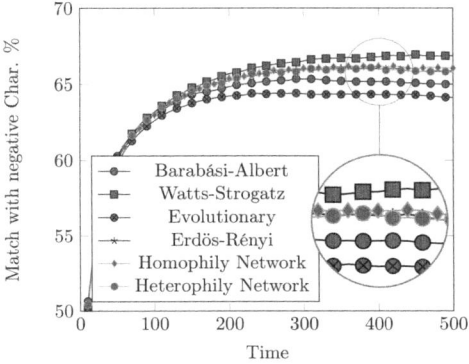

Figure 67: Match with Negative Characteristics for Different Network Topologies.

The explanation for the results is also evidenced by the data depicted in Figure 68. The figure shows that the number of firms for network topologies is following the same ranking as in the previous figure. As also explained in chapter 5.2.3, if consumers in networks affect each other's preferences the degree distribution has a major effect on the speed and degree of market's convergence towards markets with homogenous demand. The heterogeneity of markets, however, is the source of functioning markets with dynamic segmentation and, hence, with innovation.

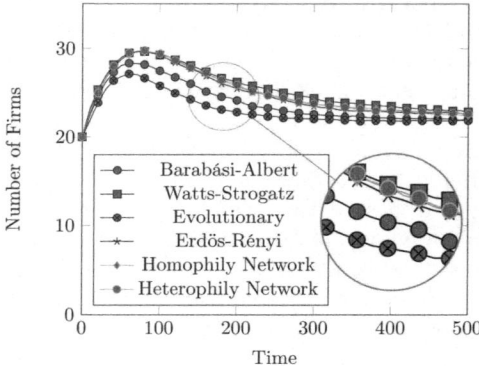

Figure 68: Number of Firms for Different Network Topologies.

6.3 Discussion

The discussion around RRI points at an important and too often neglected aspect of innovation. Although innovations generally are attributed with something always positive and desirable (Blok, Lemmens 2015), extending the scope of innovation towards a multidimensional approach, leads to a new perspective. Any innovation is multi-faceted and, thus, presents positive and negative aspects. Correspondingly, we need to consider that innovation also have a set of negative characteristics which some consumers consider and some do not. Hence, in a similar vein as in the previous chapters we see the consumer side as the driving force for the successful establishment of new and *responsible* products.

Our model allows for an *in-silicio* experiment for the analysis of the creation and diffusion of responsible innovation. Consumers not only face products with characteristics such as in the previous experiments, but also face harmful characteristics, i.e. the Faustian aspect of innovation (Blok, Lemmens 2015). This Faustian aspect of innovation covers potentially negative characteristics such as the energy consumption, the waste production involved but also criteria as work safety during production etc. Building on the baseline model from chapter 4 and the extensions made in chapter 5 we applied in this chapter different scenarios, each analysing the creation and diffusion of responsible innovation in different environments. Our results show that depending on the market environment different socially optimal market compositions emerge.

Summing up this chapter's results, we first see that markets of heterogeneous consumers produce better products (i.e. products which match with the theoretical socially optimum) than markets of homogeneous demand. In a situation of homogeneous demand, firms specialise on either matching the individual demand of consumers or on producing products which match with the socially optimum. This in turn, reduces innovation in the markets and, thus, reduces the advancements necessary to produce responsible products. Second, depending on the scarcity of information, markets sometime fall into something what we call *responsibility gap*. Markets with less informed consumers, focusing only on the negative product characteristic, potentially lead to a situation which fails to produce enough innovation to produce responsible products. Instead, a more balanced composition of consumers is preferable in which some consumers consider negative characteristics and some do not. The last experiment showed that also networks between consumers have significant effects on the creation and diffusion of responsible innovation and, thus, must be considered.

7 Discussion and Further Research Avenues

Given the current momentum of research on innovation it is surprising that something essential and vital as the demand side and the role of consumer has got so little attention. In the tradition of Schumpeter, today's research focus of evolutionary and neo-Schumpeterian economists still lies on the supply side and, thus, on firms. At the very heart of evolutionary economics lies the abandonment of restrictive assumptions and equilibrium oriented frameworks. Against this backdrop, it is surprising that many scholars in this field still oversimplify consumer's behaviour and treat them as a passive element in the complex system of innovation. Especially if we consider that the intense debate of the 1960s and 1970s already lead to an appeasing consensus in which the two contrasting perspectives of demand-pull and supply-push advocates were combined to an understanding that both factors are relevant, it is striking how little has been done to analyse the mutual relationship between both sides.

This theoretical gap cannot be seen in isolation. In fact, it is strongly related to limitations of modelling frameworks in use. To study innovation as the emerging outcome of the actions and interactions between heterogeneous economic actors we need a powerful modelling approach which can incorporate individual behaviour and individual heterogeneity. Building on this, the ABM approach is an appropriate approach to tackle this issue. ABMs designed to study demand side effects on innovation, however, are only at the beginning. So far, no ABM can offer a complete picture. Instead, we first need to establish widely accepted frameworks focusing on fundamental processes to gain a basic understanding of the relevant processes and dynamics emerging. Building on that we can stepwise increase the complexity, adding further aspects into models.

To cope with our so far limited understanding on the role of the demand side, we first need to build a simplified model of the reality. Consequently, our first aim was not to present an all-encompassing model of innovation and demand. Instead, we aimed to present a general framework model which recognises that innovations have multidimensional characteristics. With this, it is only natural that also the demand for innovations is heterogeneous (Lancaster 1966). Consequently, the first challenge was to find an adequate representation of knowledge of firms, products and consumers demand and a suitable and flexible method to transform knowledge into marketable products within an innovative environment. With this we are able to model the complex dynamics between these layers of innovation and, hence, the coevolution of demand and supply.

Although the processes introduced in the model in chapter 4 may be considered simplistic, they are nonetheless able to reproduce fundamental aspects, which have to be considered for the analysis of innovation processes. Consumer heterogeneity creates static and dynamic segmentation of markets and by that a persistent innovation dynamic beyond any equilibrium consideration without reaching a steady state. In fact, only in heterogeneous markets firms face an endogenous force to create new innovations.

The model is able to reproduce log-normal as well as normal firm size distributions. This leads to the interesting hypothesis that the particular firm size distribution in an industry simply mirrors consumers' heterogeneity.

Finally, the simulation model also allows for *in-silicio* experiments, showing the possible implications of including consumer heterogeneity into the analysis. Our results support the idea that only a sound design of innovation policies can foster industrial development. These policy measures have to consider both sides of the market carefully and evaluate which market side should be influenced by policy interventions. However, especially in the beginning of an emerging industry demand sided measures seem to be superior compared to supply side measures.

Building on this baseline model, we extended the scope of analysis in different ways. Focusing on the interactions of firms in informal knowledge networks, chapter 5.1.2 introduced a simple model of knowledge diffusion. Our investigation of a static network reveals that networks exhibit a structural characteristic which is too often overlooked: the degree distribution of agents. Our results show that the heterogeneity in degree centrality in networks hinders an efficient and fast knowledge diffusion. Especially less embedded agents are not able to keep up with the system and, hence, stop trading, hereby interrupting the knowledge flow in the whole network.

One key element of actors' strategies to link with others is preferential attachment, according to which linking up to other high degree actors is likely to increase the individual knowledge stock (Barabasi, Albert 1999, Mueller et al. 2014). Exactly this linking strategy leads to network topologies in which the big get bigger creating in total a skewed degree distribution. This, however, hinders the diffusion process, and, in turn, prevents smaller nodes from gaining knowledge. To put it another way, the limiting network characteristics, which hinder the knowledge diffusion at both the actor and the network level, are actually caused by the myopic behaviour of small economic actors which aim at receiving knowledge. In contrast to how small actors actually behave, the optimal strategy for gaining knowledge might actually be to link up with other small actors. In a final policy experiment we showed that the effect described above can lead to situations in which additional links actually have a negative effect on the network's performance.

Finally, in chapter 5.2 we introduced boundedly rational consumers and networks between consumers. Starting with implementing the concept of limitedly informed consumers we see that studies on the role of consumer must be aware of the heterogeneity of information amongst consumers. The scarcity of information creates additional heterogeneity amongst consumers and changes the way firms engage in innovation activities. Especially in fully homogeneous and fully heterogeneous markets this scarcity pays off badly and limits the market's possibilities to create products matching consumers' preferences. Additional the results show that it is also important to refer to policy measures aiming at informing consumers and building awareness.

Second, building on this scenario of scarcity of knowledge consumer networks become important sources of information. Consumer networks can reduce the information scarcity in markets and increase the overall market performance. However, the overall network topology on average may be without any major effects. Instead, our results give evidence that for consumer networks the cognitive homophily is a decisive factor.

Finally, in the experiments of chapter 6 we applied the model for an *in-silicio* experiment for the analysis of the creation and diffusion of responsible innovation. In this situation, consumers not only face products with characteristics such as in the previous experiments, but also face harmful characteristics. The results show that depending on the market environment different socially optimal market compositions emerge and under specific conditions a full focus on negative product characteristics may even be harmful.

Our aim was not to create a full picture of the economy. Instead, our aim was to introduce a general framework model of heterogeneous consumers. As a consequence, our results only give first evidence about the causal relationship at action. The model presented in this dissertation allows for a number of additional extensions. So for example, considering only consumers and firms as driving forces in the innovation process, clearly misses other important elements in the system. Universities and other research institutes offering new knowledge and technologies, the political institutions as potential buyers (via public procurement) or as legal institutions responsible for regulations are just two important examples. Additionally, the model presented considered only a fixed number of products and product characteristics. In reality a clear cut between different products, however, is often impossible (Lancaster 1966).

The results presented offer a new perspective on the innovation process. Building on the theoretical *state of the art* shown in chapter 2, it becomes clear that the demand side and consumers are important drivers of innovation, determining the market composition and structure. Neglecting their importance and influence leads to incomplete models and, to exaggerate this issue at bit, must lead to false conclusions. Understanding innovations requires that we understand the role of the demand side. In the introduction we started by stressing Alfred North Whitehead's claim that the most important invention in the 19th century was the method of invention (Whitehead 1975). What changed during the 19th century was the way science and technology was perceived and how the people of that time managed the complexity of the scientific process. In this sense, including the demand side into the analysis of innovation processes may eventually lead to a new understanding of the method of innovation.

Literature

ADNER, R. and LEVINTHAL, D., 2001. Demand heterogeneity and technology evolution: implications for product and process innovation. Management science, 47(5), pp. 611-628.

AHRWEILER, P., PYKA, A. and GILBERT, N., 2004. Simulating knowledge dynamics in innovation networks (SKIN). Leombruni R., und Richiardi M.(Eds..), "The Agent-Based Computational Approach", World Scientific Press, Singapore, pp. 284-296.

ANDERSEN, E.S., 2001. Satiation in an evolutionary model of structural economic dynamics. Journal of Evolutionary Economics, 11(1), pp. 143-164.

ANDERSEN, E.S., 2007. Innovation and demand. The Elgar Companion to Neo-Schumpeterian Economics. Cheltenham, UK: Elgar pp. 754-765.

ANDERSEN, E.S., 2013. Evolutionary economics: post-Schumpeterian contributions. Routledge.

ANTONELLI, C. and GEHRINGER, A., 2012. Knowledge externalities and demand pull: The European evidence. Economic Systems, 39.4, pp. 608-631.

ANTONELLI, C. and GEHRINGER, A., 2015. The Competent Demand Pull. In: F. CRESPI and F. QUATRARO, eds, The Economics of Knowledge, Innovation and Systemic Technology Policy. New York: Routledge.

ANTONELLI, C. and SCELLATO, G., 2008. Out of Equilibrium Profit and Innovation. U. of Torino Department of Economics Research Paper, (4).

ARMSTRONG, M., CORNUT, G., DELACÔTE, S., LENGLET, M., MILLO, Y., MUNIESA, F., POINTIER, A. and TADJEDDINE, Y., 2012. Towards a practical approach to responsible innovation in finance: New Product Committees revisited. Journal of Financial Regulation and Compliance, 20(2), pp. 147-168.

ARTHUR, W.B., 1989. Competing technologies, increasing returns, and lock-in by historical events. The economic journal, 99(394), pp. 116-131.

AXELROD, R., 1997. Advancing the art of simulation in the social sciences. In: R. CONTE, R. HEGSELMANN and P. TERNA, eds, Simulating Social Phenomena. Berlin: Springer, pp. 21-40.

AXTELL, R.L. and EPSTEIN, J.M., 1994. Agent-based modeling: understanding our creations. The Bulletin of the Santa Fe Institute, 9(2), pp. 28-32.

BABUTSIDZE, Z., 2012. if your love it - I'll probably hate it: Local interactions amon consumers of information goods. Documents de Travail de l'OFCE 2012-24, Observatoire Francais des Conjonctures Economiques (OFCE).

BABUTSIDZE, Z., 2015. Innovation, competition and firm size distribution on fragmented markets. Journal of Evolutionary Economics, pp. 1-27.

BALCONI, M., BRUSONI, S. and ORSENIGO, L., 2010. In defence of the linear model: An essay. Research Policy, 39(1), pp. 1-13.

BANERJEE, A., CHANDRASEKHAR, A.G., DUFLO, E. and JACKSON, M.O., 2013. The diffusion of microfinance. Science (New York, N.Y.), 341(6144), pp. 1236498.

BARABASI, A.L. and ALBERT, R., 1999. Emergence of scaling in random networks. Science, 286(5439), pp. 509-512.

BASS, F., 1969. A New Product Growth for Model Consumer Durables. Management Science, 15(5), pp. 215-227.

BAZGHANDI, A., 2012. Techniques, advantages and problems of agent based modeling for traffic simulation. Int J Comput Sci, 9.1, pp. 115-119.

BERG, J., DICKHAUT, J. and MCCABE, K., 1995. Trust, reciprocity, and social history. Games and Economic Behavior, 10(1), pp. 122-142.

BIANCHI, C., CIRILLO, P., GALLEGATI, M. and VAGLIASINDI, P.A., 2008. Validation in agent-based models: An investigation on the CATS model. Journal of Economic Behavior & Organization, 67(3), pp. 947-964.

BLEDA, M. and VALENTE, M., 2009. Graded eco-labels: a demand-oriented approach to reduce pollution. Technological Forecasting and Social Change, 76(4), pp. 512-524.

BLOK, V. and LEMMENS, P., 2015. The emerging concept of responsible innovation. Three reasons why it is questionable and calls for a radical transformation of the concept of innovation. Responsible Innovation 2. Springer International Publishing, pp. 19-35.

BOLLOBÁS, B., RIORDAN, O., SPENCER, J. and TUSNÁDY, G., 2001. The degree sequence of a scale free random graph process. Random Structures & Algorithms, 18(3), pp. 279-290.

BONABEAU, E., 2002. Agent-based modeling: methods and techniques for simulating human systems. Proceedings of the National Academy of Sciences of the United States of America, 99 Suppl 3, pp. 7280-7287.

BORGATTI, S.P., MEHRA, A., BRASS, D.J. and LABIANCA, G., 2009. Network analysis in the social sciences. Science, 323(5916), pp. 892-895.

BOULDING, K.E., 1991. What is evolutionary economics? Journal of Evolutionary economics, 1(1), pp. 9-17.

BOUSQUET, F. and LE PAGE, C., 2004. Multi-agent simulations and ecosystem management: a review. Ecological Modelling, 176(3-4), pp. 313-332.

BURT, R.S., 1995. Structural holes: The social structure of competition. Cambridge: Harvard University Press.

BUSH, V., 1945. Science: The endless frontier. Transactions of the Kansas Academy of Science (1903), pp. 231-264.

CABRAL, L.M. and MATA, J., 2003. On the evolution of the firm size distribution: Facts and theory. American economic review, pp. 1075-1090.

CANTNER, U. and HANUSCH, H., 1999. Heterogeneity and Evolutionary Change.

CANTNER, U. and HANUSCH, H., 2001. Heterogeneity and Evolutionary Change: Empirical Conception, Findings, and Unresolved Issues. Frontiers of Evolutionary Economics, edited by John Foster and J.Stan Metcalfe, pp. 228-267.

CANTONO, S. and SILVERBERG, G., 2009. A percolation model of eco-innovation diffusion: the relationship between diffusion, learning economies and subsidies. Technological forecasting and social change, 76(4), pp. 487-496.

CAPLAT, P., ANAND, M. and BAUCH, C., 2008. Symmetric competition causes population oscillations in an individual-based model of forest dynamics. Ecological Modelling, 211(3–4), pp. 491-500.

CHAMBERLIN, E.H., 1933. The theory of monopolistic competition. Vol. 6 edn. Cambridge: MA: Harvard University Press.

CIARLI, T., LORENTZ, A., SAVONA, M. and VALENTE, M., 2008. Structural change of production and consumption: a micro to macro approach to economic growth and income distribution. LEM Working Paper Series, (No. 2008/08).

CIARLI, T., LORENTZ, A., SAVONA, M. and VALENTE, M., 2010. The effect of consumption and production structure on growth and distribution. A micro to macro model. Metroeconomica, 61(1), pp. 180-218.

COLEMAN, J.S., 1988. Social capital in the creation of human capital. American journal of sociology, 94(S1), pp. 95-120.

COOMBS, R., SAVIOTTI, P. and WALSH, V., 1987. Economics and technological change. Rowman & Littlefield.

COOMBS, R., 2001. Technology and the market: demand, users and innovation. Edward Elgar Publishing.

COWAN, R. and JONARD, N., 2007. Structural holes, innovation and the distribution of ideas. Journal of Economic Interaction and Coordination, 2(2), pp. 93-110.

COWAN, R., COWAN, W. and SWANN, P., 1997. A model of demand with interactions among consumers. International Journal of Industrial Organization, 15(6), pp. 711-732.

COWAN, R. and JONARD, N., 2004. Network structure and the diffusion of knowledge. Journal of Economic Dynamics and Control, 28(8), pp. 1557-1575.

DAS, S., 2006. On agent-based modeling of complex systems: Learning and bounded rationality. Department of Computer Science and Engineering. La Jolla, CA, pp. 92093-90404.

DAVID, P.A., 1985. Clio and the Economics of QWERTY. The American Economic Review, 75(2), pp. 332-337.

DAWID, H. and KOPEL, M., 1998. On economic applications of the genetic algorithm: a model of the cobweb type. Journal of Evolutionary Economics, 8(3), pp. 297-315.

DIXIT, A.K. and STIGLITZ, J.E., 1977. Monopolistic competition and optimum product diversity. The American Economic Review, 67(3), pp. 297-308.

DORAN, J., 2001. Intervening to achieve co-operative ecosystem management: towards an agent based model. Journal of Artificial Societies and Social Simulation, 4(2), pp. 1-21.

DOSI, G., 1982. Technological paradigms and technological trajectories: a suggested interpretation of the determinants and directions of technical change. Research policy, 11(3), pp. 147-162.

DOSI, G., 1988a. The nature of the innovative process. In: G. DOSI, C. FREEMAN, R. NELSON, G. SILVERBERG and L. SOETE, eds, Technical Change and Economic Theory. London, pp. 590-607.

DOSI, G., 1988b. Sources, procedures, and microeconomic effects of innovation. Journal of economic literature, 26(3), pp. 1120-1171.

DOSI, G. and NELSON, R.R., 1994. An introduction to evolutionary theories in economics. Journal of evolutionary economics, 4(3), pp. 153-172.

DOWNING, T.E., MOSS, S. and PAHL-WOSTL, C., 2001. Understanding climate policy using participatory agent-based social simulation. Multi-Agent-Based Simulation. Springer, pp. 198-213.

DRIESCH, H., 1908. The science and philosophy of the organism. Aberdeen: Aberdeen University Press.

DUGUNDJI, E.R. and GULYÁS, L., 2008. Sociodynamic discrete choice on networks in space: impacts of agent heterogeneity on emergent outcomes. Environment and planning. B, Planning & design, 35(6), pp. 1028.

EDGERTON, D., 2004. The Linear Model. Did not Exist: Reflections of the History and Historiography of Science and Research in Industry in the Twentieth Century. In The science-industry nexus: history, policy, implications. Massachusetts, USA: Science History Publications.

EDLER, J., 2007. Bedürfnisse als Innovationsmotor: Konzepte und Instrumente nachfrageorientierter Innovationspolitik. Berlin: sigma.

EDLER, J. and GEORGHIOU, L., 2007. Public procurement and innovation—Resurrecting the demand side. Research policy, 36(7), pp. 949-963.

EDLER, J., 2009. Demand policies for innovation. Manchester Business School Research Paper, (579).

EDMONDS, B., 1999. Modelling bounded rationality in agent-based simulations using the evolution of mental models. Computational techniques for modelling learning in economics. Springer, pp. 305-332.

EDMONDS, B. and MOSS, S., 2005. From KISS to KIDS–an 'anti-simplistic'modelling approach. Springer.

EDMONDS, B., 2001. The use of models-making MABS more informative. Multi-agent-based simulation, pp. 269-282.

EDQUIST, C., 1994. Technology policy: the interaction between governments and markets. DE GRUYTER STUDIES IN ORGANIZATION, pp. 67-67.

ELIASSON, G., 1991. Modeling the experimentally organized economy: Complex dynamics in an empirical micro-macro model of endogenous economic growth. Journal of Economic Behavior & Organization, 16(1), pp. 153-182.

EPSTEIN, J.M., 1999. Agent-based computational models and generative social science. Generative Social Science: Studies in Agent-Based Computational Modeling, 4(5), pp. 4-46.

EPSTEIN, J.M., 2006. Generative social science: Studies in agent-based computational modeling. Princeton University Press.

EPSTEIN, J.M. and AXTELL, R., 1996. Growing artificial societies: Social Science from the Bottom Up. Washington: Brookings Institution Press.

ERDŐS, P. and RÉNYI, A., 1959. On random graphs. Publicationes Mathematicae Debrecen, 6, pp. 290-297.

ERDŐS, P. and RÉNYI, A., 1960. On the evolution of random graphs. Publications of the Mathematical Institute of the Hungarian Academy of Sciences, No.5, pp. 17-61.

EUROPEAN COMMISION, 2016. Why do we need an Innovation Union? http://ec.europa.eu/research/innovation-union/index_en.cfm?pg=why [February 22, 2016].

EVANS, D.S., 1987. Tests of alternative theories of firm growth. The journal of political economy, pp. 657-674.

FABER, A. and FRENKEN, K., 2009. Models in evolutionary economics and environmental policy: Towards an evolutionary environmental economics. Technological Forecasting and Social Change, 76(4), pp. 462-470.

FAGERBERG, J., 2003. Schumpeter and the revival of evolutionary economics: an appraisal of the literature. Journal of evolutionary economics, 13(2), pp. 125-159.

FAGERBERG, J., 2004. Innovation: a guide to the literature. In: J. FAGERBERG, J. MOWERY and R.R. NELSON, eds, The Oxford Handbook of Innovation. Georgia Institute of Technology.

FAGIOLO, G. and ROVENTINI, A., 2012. Macroeconomic policy in dsge and agent-based models. Revue de l'OFCE, 124(5), pp. 67-116.

FARMER, J.D. and FOLEY, D., 2009. The economy needs agent-based modelling. Nature, 460(7256), pp. 685-686.

FERBER, J., 1995. Multi-agent systems: an introduction to distributed artificial intelligence. Addison-Wesley.

FLAKE, G.W., 1998. The computational beauty of nature: Computer explorations of fractals, chaos, complex systems, and adaptation. MIT press.

FLEMING, L., 2001. Recombinant uncertainty in technological search. Management science, 47(1), pp. 117-132.

FLEMING, L. and SORENSON, O., 2001. Technology as a complex adaptive system: evidence from patent data. Research Policy, 30(7), pp. 1019-1039.

FOGLI, A. and VELDKAMP, L., 2014. Germs, Social Networks and Growth.

FREEMAN, L.C., 2004. The development of social network analysis. Empirical Press Vancouver, British Columbia.

FRENKEN, K.K., 2004. History, state and prospects of evolutionary models of technical change: a review with special emphasis on complexity theory. Complexity focus.

FRENKEN, K., 2006. Innovation, evolution and complexity theory. Edward Elgar Publishing.

FRIEDMAN, M., 1953. The Methodology of Positive Economics. In: M. FRIEDMAN, ed, Essays in positive economics. Third Edition edn. University of Chicago Press, pp. 3-43.

GALLOUJ, F. and WEINSTEIN, O., 1997. Innovation in services. Research policy, 26(4), pp. 537-556.

GEIGER, N., 2015. Wellen wirtschaftlichen Wandels–theoretische, historische und statistische Betrachtung.

GELL-MANN, M., 1995. The Quark and the Jaguar: Adventures in the Simple and the Complex. Macmillan.

GIBBONS, M. and GUMMETT, P., 1977. Recent Changes in Government Administration of Research and Development: A New Context for Innovation, International Symposium on Industrial Innovation, Strathclyde University, September 1977.

GIBBONS, M. and JOHNSTON, R., 1974. The roles of science in technological innovation. Research Policy, 3(3), pp. 220-242.

GIBRAT, R., 1931. Les inégalités économiques: applications: aux inégalités des richesses, à la concentration des entreprises, aux populations des villes, aux statistiques des familles, etc: d'une loi nouvelle: la loi de l'effet proportionnel. Librairie du Recueil Sirey.

GIGERENZER, G. and SELTEN, R., 2002. Bounded rationality: The adaptive toolbox. Mit Press.

GILBERT, N., 2008. Agent-based models. Sage.

GILBERT, N., 1997-last update, A simulation of the structure of academic science [Homepage of Sociological Research Online], [Online]. Available: http://www.socresonline.org.uk/socresonline/2/2/3.html.

GILBERT, N. and TERNA, P., 2000. How to build and use agent-based models in social science. Mind & Society, 1(1), pp. 57-72.

GILBERT, N., PYKA, A. and AHRWEILER, P., 2001. Innovation networks-a simulation approach. Journal of Artificial Societies and Social Simulation, 4(3), pp. 1-14.

GILBERT, N. and TROITZSCH, K., 2005. Simulation for the social scientist. Open Univ Pr.

GILBERT, N., AHRWEILER, P. and PYKA, A., 2007. Learning in innovation networks: Some simulation experiments. Physica A: Statistical Mechanics and its Applications, 378(1), pp. 100-109.

GODIN, B., 2006. The Linear model of innovation the historical construction of an analytical framework. Science, Technology & Human Values, 31(6), pp. 639-667.

GODIN, B. and LANE, J., 2013. Pushes and Pulls Hi (S) tory of the Demand Pull Model of Innovation. Science, Technology & Human Values, 38(5), pp. 621-654.

GÜTH, W., SCHMITTBERGER, R. and SCHWARZE, B., 1982. An experimental analysis of ultimatum bargaining. Journal of economic behavior & organization, 3(4), pp. 367-388.

HALL, B.H., 1987. The relationship between firm size and firm growth in the US manufacturing sector. Journal of Industrial Economics, 35(4), pp. 583-606.

HAMMING, R.W., 1950. Error detecting and error correcting codes. Bell System technical journal, 29(2), pp. 147-160.

HANUSCH, H. and PYKA, A., 2007a. Principles of neo-Schumpeterian economics. Cambridge Journal of Economics, 31(2), pp. 275-289.

HANUSCH, H. and PYKA, A., 2007b. Schumpeter, Joseph Alois (1883–1950). Elgar Companion to Neo-Schumpeterian Economics}. Edward Elgar, pp. 19-27.

HANUSCH, H. and PYKA, A., 2007c. Manifesto for Comprehensive Neo-Schumpeterian Economics. History of Economic Ideas, 15(1), pp. 23-42.

HARE, M. and DEADMAN, P., 2004. Further towards a taxonomy of agent-based simulation models in environmental management. Mathematics and Computers in Simulation, 64(1), pp. 25-40.

HARVEY, M., MCMEEKIN, A., RANDLES, S., SOUTHERTON, D., TETHER, B. and WARDE, A., 2001. Between demand & consumption: A framework for research. Centre for Research on Innovation and Competition, University of Manchester.

HELLSTRÖM, T., 2003. Systemic innovation and risk: technology assessment and the challenge of responsible innovation. Technology in Society, 25(3), pp. 369-384.

HERSTAD, S.J., SANDVEN, T. and SOLBERG, E., 2013. Location, education and enterprise growth. Applied Economics Letters, 20(10), pp. 1019-1022.

HODGSON, G.M., 1993. Economics and evolution: bringing life back into economics. University of Michigan Press.

HODGSON, G.M., 1997. The evolutionary and non-Darwinian economics of Joseph Schumpeter. Journal of Evolutionary Economics, 7(2), pp. 131-145.

HODGSON, G.M., 1998a. The approach of institutional economics. Journal of Economic Literature, 36(1), pp. 166-192.

HODGSON, G.M., 1998b. On the evolution of Thorstein Veblen's evolutionary economics. Cambridge Journal of Economics, 22(4), pp. 415-431.

HODGSON, G.M., 1999. Evolution and institutions. Cheltenham: Edward Elgar.

HOLLAND, J.H., 1995. Hidden order: How adaptation builds complexity. Reading: Addison-Wesley.

HOTELLING, H., 1929. Stability in Competition. The Economic Journal, 39(153), pp. 41-57.

JENNINGS, S., 1998. Wooldridge. A Roadmap of Agent Research and Development (Autonomous Agents and Multi-Agent Systems).

JUDSON, O.P., 1994. The rise of the individual-based model in ecology. Trends in Ecology & Evolution, 9(1), pp. 9-14.

KAHNEMAN, D., SLOVIC, P. and TVERSKY, A., 1974. Judgment under uncertainty: Heuristics and biases. Science, 185.4157, pp. 1124-1131.

KAHNEMAN, D. and TVERSKY, A., 1979. Prospect theory: An analysis of decision under risk. Econometrica: Journal of the Econometric Society, pp. 263-291.

KALDOR, N., 1966. Causes of the Slow Rate of Growth in the United Kingdom. Cambridge (Mass.), Cambridge University Press.

KALDOR, N., 1972. The irrelevance of equilibrium economics. The Economic Journal, pp. 1237-1255.

KALDOR, N., 1975. Economic growth and the Verdoorn Law--A Comment on Mr Rowthorn's Article. The Economic Journal, pp. 891-896.

KAUFFMAN, S. and LEVIN, S., 1987. Towards a general theory of adaptive walks on rugged landscapes. Journal of theoretical biology, 128(1), pp. 11-45.

KAUFFMAN, S.A. and WEINBERGER, E.D., 1989. The NK model of rugged fitness landscapes and its application to maturation of the immune response. Journal of theoretical biology, 141(2), pp. 211-245.

KEYNES, J.M., 1937. The general theory of employment. The Quarterly Journal of Economics, 51(2), pp. 209-223.

KIESLING, E., GÜNTHER, M., STUMMER, C. and WAKOLBINGER, L.M., 2012. Agent-based simulation of innovation diffusion: a review. Central European Journal of Operations Research, 20(2), pp. 183-230.

KLEINKNECHT, A. and VERSPAGEN, B., 1990. Demand and innovation: Schmookler re-examined. Research policy, 19(4), pp. 387-394.

KLINE, S.J., 1985. Innovation is not a linear process. Research management, 28(4), pp. 36-45.

KLINE, S.J. and ROSENBERG, N., 1986. An overview of innovation. The positive sum strategy: Harnessing technology for economic growth, 14, pp. 640.

KNELL, M., 2012. Demand driven innovation in Economic Thought, UNDERPINN conference materials: Demand, Innovation and Policy, Manchester Institute of Innovation Research, MBS, University of Manchester Lall S. (2000), Foreign direct investment, technology development and competitiveness: issues and evidence [in:] 2012.

KNIGHT, F.H., 1921. Risk, Uncertainty and Profit. Boston: Hart, Schaffner & Marx.

KUDIC, M., 2014. Innovation Networks in the German Laser Industry: Evolutionary Change, Strategic Positioning, and Firm Innovativeness. Springer.

KURZ, H.D., 2005. Joseph A. Schumpeter: ein Sozialökonom zwischen Marx und Walras. Metropolis-Verlag GmbH.

KWASNICKI, W., 2007. 25 Schumpeterian modelling. Elgar Companion to Neo-Schumpeterian Economics, pp. 389.

LABARBERA, M., 1983. Why the wheels won't go. American Naturalist, pp. 395-408.

LANCASTER, K., 1975. Socially optimal product differentiation. The American Economic Review, 65(4), pp. 567-585.

LANCASTER, K., 1979. Variety, equity, and efficiency: product variety in an industrial society. New York: Columbia University Press.

LANCASTER, K.J., 1966. A new approach to consumer theory. The journal of political economy, pp. 132-157.

LANGRISH, J., GIBBONS, M., EVANS, W.G. and JEVONS, F.R., 1972. Wealth from knowledge: studies of innovation in industry. London: Macmillan.

LEOMBRUNI, R., 2002. The methodological status of agent-based simulations. LABORatorio Riccardo Revelli-Centre for Employment Studies Working Paper, (19).

LERMAN, K., YAN, X. and WU, X., 2015. The majority illusion in social networks. arXiv preprint arXiv:1506.03022.

LEVÉN, P., HOLMSTRÖM, J. and MATHIASSEN, L., 2014. Managing research and innovation networks: Evidence from a government sponsored cross-industry program. Research Policy, 43(1), pp. 156-168.

LEVINTHAL, D.A., 1997. Adaptation on rugged landscapes. Management science, 43(7), pp. 934-950.

LIN, M. and LI, N., 2010. Scale-free network provides an optimal pattern for knowledge transfer. Physica A: Statistical Mechanics and its Applications, 389(3), pp. 473-480.

LORENTZ, A., CIARLI, T., SAVONA, M. and VALENTE, M., 2015. The effect of demand-driven structural transformations on growth and technological change. Journal of Evolutionary Economics, pp. 1-28.

LUCAS JR, R.E., 1978. On the size distribution of business firms. The Bell Journal of Economics, pp. 508-523.

MACAL, C.M. and NORTH, M.J., 2005. Tutorial on agent-based modeling and simulation, Proceedings of the 37th conference on Winter simulation 2005, Winter Simulation Conference, pp. 2-15.

MALERBA, F., 1992. Learning by firms and incremental technical change. The economic journal, 102(413), pp. 845-859.

MALERBA, F., 2007. Innovation and the dynamics and evolution of industries: Progress and challenges. International Journal of Industrial Organization, 25(4), pp. 675-699.

MARSHALL, A., 1890. The Principles of Economics. London: Macmillan.

MAYR, E., 1959. Darwin and the evolutionary theory in biology. Evolution and anthropology: A centennial appraisal, pp. 1-10.

METCALFE, J.S., 2001. Consumption, preferences, and the evolutionary agenda. Springer.

MILLING, P.M., 1996. Modeling innovation processes for decision support and management simulation. System Dynamics Review, 12(3), pp. 211-234.

MILLING, P.M., 2002. Understanding and managing innovation processes. System Dynamics Review, 18(1), pp. 73-86.

MINSKY, M., 1965. Matter, mind and models. Proceedings of the IFID Congress.

MISES, L.V. and ACTION, H., 1949. A Treatise on Economics. Yale University.

MOORE, G., 1965. Cramming More Components Onto Integrated Circuits. Electronics,(38) 8.

MORONE, P. and TAYLOR, R., 2010. Knowledge diffusion and innovation: Modelling complex entrepreneurial behaviours. Edward Elgar Publishing.

MORONE, A., MORONE, P. and TAYLOR, R., 2007. A laboratory experiment of knowledge diffusion dynamics. In: U. CANTER and F. MALERBA, eds, Innovation, Industrial Dynamics and Structural Transformation. Heidelberg: Springer, pp. 283-302.

MOSS, S. and EDMONDS, B., 2005. Sociology and Simulation: Statistical and Qualitative Cross-Validation. American Journal of Sociology, 110(4), pp. 1095-1131.

MOWERY, D. and ROSENBERG, N., 1979. The influence of market demand upon innovation: a critical review of some recent empirical studies. Research policy, 8(2), pp. 102-153.

MUELLER, M., BUCHMANN, T. and KUDIC, M., 2014. Micro Strategies and Macro Patterns in the Evolution of Innovation Networks – An Agent-Based Simulation Approach. In: N. GILBERT, P. AHRWEILER and A. PYKA, eds, Simulating knowledge dynamics in innovation networks. Heidelberg: Springer, pp. 73-95.

MUELLER, M., SCHREMPF, B. and PYKA, A., 2015. Simulating demand-side effects on innovation. International Journal of Computational Economics and Econometrics, 5(3), pp. 220-236.

MYERS, S. and MARQUIS, D.G., 1969. Successful industrial innovation. Institute of Public Administration.

NELSON, R.R. and WINTER, S.G., 1973. Toward an evolutionary theory of economic capabilities. The American Economic Review, pp. 440-449.

NELSON, R.R. and WINTER, S.G., 1974. Neoclassical vs. evolutionary theories of economic growth: critique and prospectus. The Economic Journal, pp. 886-905.

NELSON, R.R. and WINTER, S.G., 1975. Growth theory from an evolutionary perspective: The differential productivity puzzle. The American Economic Review, pp. 338-344.

NELSON, R.R. and WINTER, S.G., 1977. Simulation of Schumpeterian competition. The American Economic Review, 67(1), pp. 271-276.

NELSON, R.R. and CONSOLI, D., 2010. An evolutionary theory of household consumption behavior. Journal of Evolutionary Economics, 20(5), pp. 665-687.

NELSON, R. and WINTER, S., 1982. An evolutionary theory of economic change.

NEWBY, H., 1992. One society, one Wissenschaft: a 21st century vision. Science and Public Policy, 19(1), pp. 7-14.

OECD, 1996. The Knowledge-based Economy. General Distribution OCDE/GD, 96(102).

ORCUTT, G.H., MERZ, J. and QUINKE, H., 1986. Microanalytic simulation models to support social and financial policy. North-Holland Amsterdam:.

ORMEROD, P. and ROSEWELL, B., 2009. Validation and verification of agent-based models in the social sciences. Epistemological Aspects of Computer Simulation in the Social Sciences. Springer, pp. 130-140.

OWEN, R., BAXTER, D., MAYNARD, T. and DEPLEDGE, M., 2009. Beyond Regulation: Risk Pricing and Responsible Innovation†. Environmental science & technology, 43(18), pp. 6902-6906.

OWEN, R. and GOLDBERG, N., 2010. Responsible innovation: a pilot study with the UK Engineering and Physical Sciences Research Council. Risk Analysis, 30(11), pp. 1699-1707.

PAVITT, K., ROBSON, M. and TOWNSEND, J., 1987. The size distribution of innovating firms in the UK: 1945-1983. The Journal of Industrial Economics, pp. 297-316.

PEIRCE, C., 1901. Collected papers of Charles Sanders Peirce (CP 7.218—1901, On the logic of drawing history from ancient documents especially from testimonies).

PINCH, T.J. and BIJKER, W.E., 1987. The social construction of facts and artifacts. In: W.E. BIJKER, T.P. HUGHES and T. PINCH, eds, The Social Construction of Technological Systems.

PODOLNY, J.M., 2001. Networks as the Pipes and Prisms of the Market1. American journal of sociology, 107(1), pp. 33-60.

POLHILL, J.G., GOTTS, N.M. and LAW, A., 2001. Imitative versus nonimitative strategies in a land-use simulation. Cybernetics & Systems, 32(1-2), pp. 285-307.

POPPER, K., 2014. Conjectures and refutations: The growth of scientific knowledge. Routledge.

POTTS, J., 2003. Evolutionary Economics: An Introduction to the Foundation of Liberal Economic Philosophy.

POWELL, W.W. and GRODAL, S., 2005. Networks of innovators. In: J. FAGERBERG, D.C. MOWERY and R.R. NELSON, eds, The Oxford Handbook of Innovation. Oxford; New York: Oxford University Press, pp. 56-85.

PRICE, W.J. and BASS, L.W., 1969. Scientific research and the innovative process. Science (New York, N.Y.), 164(3881), pp. 802-806.

PYKA, A., 1997. Informal networking. Technovation, 17(4), pp. 207-220.

PYKA, A. and FAGIOLO, G., 2007. Agent-based Modelling: A Methodology for Neo-Schumpeterian Economics «, in: Hanusch, H. and A. Pyka (eds.), The Elgar Companion to Neo-Schumpeterian Economics, Cheltenham, Edward Elgar.

RAHMANDAD, H. and STERMAN, J., 2008. Heterogeneity and network structure in the dynamics of diffusion: Comparing agent-based and differential equation models. Management Science, 54(5), pp. 998-1014.

REYNOLDS, C.W., 1987. Flocks, herds and schools: A distributed behavioral model, ACM SIGGRAPH Computer Graphics 1987, ACM, pp. 25-34.

RICHIARDI, M.G., LEOMBRUNI, R., SAAM, N.J. and SONNESSA, M., 2006. A common protocol for agent-based social simulation. Journal of artificial societies and social simulation, 9.

ROGERS, E.M., 2010. Diffusion of innovations. Simon and Schuster.

ROSENBERG, N., 1974. Science, invention and economic growth. The Economic Journal, 84(333), pp. 90-108.

ROTHWELL, R., 1977. The characteristics of successful innovators and technically progressive firms (with some comments on innovation research). R&D Management, 7(3), pp. 191-206.

ROTHWELL, R. and ZEGVELD, W., 1985. Reindustrialization and technology. ME Sharpe.

ROTHWELL, R., 1992. Successful industrial innovation: critical factors for the 1990s. R&D Management, 22(3), pp. 221-240.

ROTHWELL, R., 1994. Towards the fifth-generation innovation process. International marketing review, 11(1), pp. 7-31.

RYAN, B. and GROSS, N.C., 1943. The diffusion of hybrid seed corn in two Iowa communities. Rural Sociology, 8(1), pp. 15.

SAVIOTTI, P. and PYKA, A., 2013a. From necessities to imaginary worlds: Structural change, product quality and economic development. Technological Forecasting and Social Change, 80(8), pp. 1499-1512.

SAVIOTTI, P.P. and METCALFE, J.S., 1984. A theoretical approach to the construction of technological output indicators. Research Policy, 13(3), pp. 141-151.

SAVIOTTI, P.P., 1996. Technological evolution, variety, and the economy. Edward Elgar Publishing.

SAVIOTTI, P.P., 2001. Variety, growth and demand. Journal of Evolutionary economics, 11(1), pp. 119-142.

SAVIOTTI, P.P. and PYKA, A., 2004. Economic development by the creation of new sectors. Journal of evolutionary economics, 14(1), pp. 1-35.

SAVIOTTI, P.P. and PYKA, A., 2012. On the co-evolution of innovation and demand: Some policy implications. Revue de l'OFCE, (5), pp. 347-388.

SAVIOTTI, P.P. and PYKA, A., 2013b. The co-evolution of innovation, demand and growth. Economics of Innovation and New technology, 22(5), pp. 461-482.

SCHELLING, T.C., 1969. Models of segregation. The American Economic Review, pp. 488-493.

SCHELLING, T.C., 1971. Dynamic models of segregation†. Journal of mathematical sociology, 1(2), pp. 143-186.

SCHERER, F.M., 1982. Demand-pull and technological invention: Schmookler revisted. The Journal of Industrial Economics, pp. 225-237.

SCHMOOKLER, J., 1962. Economic sources of inventive activity. The Journal of Economic History, 22(01), pp. 1-20.

SCHMOOKLER, J., 1966. Invention and economic growth. Harvard University Press Cambridge, MA.

SCHUMPETER, J.A., 1912. Theorie der wirtschaftlichen Entwicklung. Leipzig: Duncker & Humblot. English translation published in 1934 as The theory of economic development.

SCHUMPETER, J.A., 1928. The instability of capitalism. The economic journal, 38(151), pp. 361-386.

SCHUMPETER, J.A., 1942. Capitalism, Socialism and Democracy. New York: Harper.

SHAPIRO, F.R., 1987. Etymology of the computer bug: history and folklore. American Speech, 62(4), pp. 376-378.

SHERWIN, C.W. and ISENSON, R.S., 1967. Project hindsight. A Defense Department study of the utility of research. Science (New York, N.Y.), 156(3782), pp. 1571-1577.

SIMON, H.A., 1955. A behavioral model of rational choice. The quarterly journal of economics, pp. 99-118.

SIMON, H.A., 1957. Models of man; social and rational.

SIMON, H.A., 1959. Theories of decision-making in economics and behavioral science. The American Economic Review, 49(3), pp. 253-283.

SIMON, H.A., 1972. Theories of bounded rationality. Decision and organization, 1(1), pp. 161-176.

SIMON, H.A., 1995. Near Decomposability and Complexity: How a Mind Resides in a Brain. In: H. MOROWITZ and J. SINGER, eds, The mind, the brain and complex adaptive systems. SFI studies in the sciences of complexity, Vol. XXII. Reading: Addison-Wesley, pp. 25-43.

SMITH, A., 1776. An Inquiry into the Nature and Causes of the Wealth of Nations. Edwin Cannan's annotated edition.

SOLOW, R.M., 1956. A contribution to the theory of economic growth. The Quarterly Journal of Economics, 70(1), pp. 65-94.

SOLOW, R.M., 1957. Technical change and the aggregate production function. The review of economics and statistics, pp. 312-320.

STARFIELD, A.M., 1990. Qualitative, rule-based modeling. Bioscience, pp. 601-604.

STILGOE, J., OWEN, R. and MACNAGHTEN, P., 2013. Developing a framework for responsible innovation. Research Policy, 42(9), pp. 1568-1580.

SWANN, G.P., 1999. An economic analysis of taste-a review of Gary S. Becker: Accounting for tastes. International Journal of the Economics of Business, 6(2), pp. 281-296.

TAIT, J. and WILLIAMS, R., 1999. Policy approaches to research and development: foresight, framework and competitiveness. Science and Public Policy, 26(2), pp. 101-112.

TANG, J., FERNANDEZ-GARCIA, I., VIJAYAKUMAR, S., MARTINEZ-RUIS, H., ILLA-BOCHACA, I., NGUYEN, D.H., MAO, J., COSTES, S.V. and BARCELLOS-HOFF, M.H., 2014. Irradiation of Juvenile, but not Adult, Mammary Gland Increases Stem Cell Self-Renewal and Estrogen Receptor Negative Tumors. STEM CELLS, 32(3), pp. 649-661.

TEECE, D.J., 1986. Profiting from technological innovation: Implications for integration, collaboration, licensing and public policy. Research policy, 15(6), pp. 285-305.

TESFATSION, L., 2002. Economic agents and markets as emergent phenomena. Proceedings of the National Academy of Sciences of the United States of America, 99 Suppl 3, pp. 7191-7192.

TESFATSION, L., 2006. Agent-based computational economics: A constructive approach to economic theory. Handbook of computational economics, 2, pp. 831-880.

TVERSKY, A. and KAHNEMAN, D., 1981. The framing of decisions and the psychology of choice. Science (New York, N.Y.), 211(4481), pp. 453-458.

VALENTE, M., 1999. Evolutionary Economics and Computer Simulations.

VALENTE, M., 2008a. Laboratory for Simulation Development: LSD. LEM Working Paper Series., (No. 2008/12).

VALENTE, M., 2008b. Pseudo-NK: An enhanced model of complexity. LEM Working Paper Series.

VALENTE, M., 2009. Markets for heterogeneous products: A boundedly rational consumer model. LEM Papers Series 2009/11.

VALENTE, M., 2012. Evolutionary demand: a model for boundedly rational consumers. Journal of Evolutionary Economics, 22(5), pp. 1029-1080.

VALENTE, T.W. and ROGERS, E.M., 1995. The origins and development of the diffusion of innovations paradigm as an example of scientific growth. Science communication, 16(3), pp. 242-273.

VEBLEN, T., 1898. Why is Economics not an Evolutionary Science? The Quarterly Journal of Economics, 12(4), pp. 373-397.

VON BERTALANFFY, L., 1969. General systems theory and psychiatry–an overview. General systems theory and psychiatry, pp. 33-46.

VON HIPPEL, E., 1976. The dominant role of users in the scientific instrument innovation process. Research policy, 5(3), pp. 212-239.

VON HIPPEL, E., 1988. The sources of innovation. Oxford: Oxford University Press.

VON HIPPEL, E., 2005. Democratizing innovation: The evolving phenomenon of user innovation. Journal für Betriebswirtschaft, 55(1), pp. 63-78.

VON SCHOMBERG, R., 2007. From the ethics of technology towards and ethics of knowledge policy. Working document of the Service of the European Commission.

VON SCHOMBERG, R., 2013. A vision of responsible research and innovation. Responsible innovation: managing the responsible emergence of science and innovation in society, pp. 51-74.

WALSH, V., 1984. Invention and Innovation in the Chemical Industry: Demand-pull or Discovery-push? Research policy, 13(4), pp. 211-234.

WATTS, D.J. and STROGATZ, S.H., 1998. Collective dynamics of 'small-world'networks. Nature, 393, pp. 440-442.

WEJNERT, B., 2002. Integrating models of diffusion of innovations: A conceptual framework. Annual review of sociology, pp. 297-326.

WERKER, C. and BRENNER, T., 2004. Empirical calibration of simulation models. Papers on Economics and Evolution, .

WHITEHEAD, A., 1975. Science and the modern world: Lowell lectures, 1925.

WILENSKY, U., 1997. NetLogo Traffic Basic model. Northwestern University, Evanston, IL: Center for Connected Learning and Computer-Based Modeling.

WILENSKY, U., 1999. NetLogo. Northwestern University, Evanston, IL.: Center for Connected Learning and Computer-Based Modeling.

WILENSKY, U. and RAND, W., 2015. An introduction to agent-based modeling: modeling natural, social, and engineered complex systems with NetLogo. MIT Press.

WINDRUM, P., FAGIOLO, G. and MONETA, A., 2007. Empirical validation of agent-based models: Alternatives and prospects. Journal of Artificial Societies and Social Simulation, 10(2), pp. 8.

WITT, U., 1993. Evolutionary economics. Edward Elgar.

WITT, U., 2001a. Learning to consume–A theory of wants and the growth of demand. Journal of Evolutionary Economics, 11(1), pp. 23-36.

WITT, U., 2001b. Learning to consume–A theory of wants and the growth of demand. Journal of Evolutionary Economics, 11(1), pp. 23-36.

WITT, U., 2002. How evolutionary is Schumpeter's theory of economic development? Industry and Innovation, 9(1-2), pp. 7-22.

WITT, U., 2008. What is specific about evolutionary economics? Journal of Evolutionary Economics, 18(5), pp. 547-575.

WOLFRAM, S., 1986. Theory and applications of cellular automata. World Scientific Singapore.

WOOLDRIDGE, M. and JENNINGS, N.R., 1995. Intelligent agents: Theory and practice. The knowledge engineering review, 10(2), pp. 115-152.

ZEPPINI, P. and FRENKEN, K., 2015. Networks, Percolation, and Demand.

ZIMAN, J., 1991. A neural net model of innovation. Science and public policy, 18(1), pp. 65-75.